PROCESS EQUIPMENT SERIES

VOLUME 2

Heat Transfer Equipment

EDITED BY

Mahesh V. Bhatia, P.E.
Paul N. Cheremisinoff, P.E.

CONTRIBUTORS TO VOLUME 2

A. Cooper
Jack F. Curran
Richard C. Davis
Joe Ben Dickey, Jr.
Alan G. Furnish
D. F. Fijas
J. H. Kissel
Richard G. Krueger
Paul E. Minton
Clarence T. Moss
Michael W. Oliver

TECHNOMIC
PUBLISHING CO., INC.
LANCASTER · BASEL

Published in the Western Hemisphere by
Technomic Publishing Company, Inc.
851 New Holland Avenue
Box 3535
Lancaster, Pennsylvania 17604 U.S.A.

Distributed in the Rest of the World by
Technomic Publishing AG

Printed in the United States of America
10 9 8 7 6 5 4 3 2

Main entry under title:
    Process Equipment Series, Volume 2—Heat Transfer Equipment

A Technomic Publishing Company book
Bibliography: p.
Includes index p. 229

Library of Congress Card No. 79-63114
ISBN No. 87762-283-3

FOREWORD

This is the second volume in the *Process Equipment Series* which is a presentation of exact and useful information relating to equipment and devices used in the chemical and related process industries. The editorial material has been prepared and written by specialists in their respective fields.

The equipment described in this series is used by more than 10,000 chemical manufacturing and processing firms, employing 1.1 million Americans, with sales of approximately 100 billion dollars. Chemical industry products represent 7.5 percent of the U.S. Gross National Product (GNP) and enjoy an annual growth rate nearly twice that of the overall GNP.

Equipment, apparatus and devices employed by the process industries are so numerous and varied that the problem was to reduce such information to a moderate and practical size. The editors have given preference to equipment that has general application, and diverse uses in the process industries. The presentation has been divided among a series of volumes.

Material presented by authors is primarily written in general technical language and the text liberally supplemented with drawings and photographs. The intent has been to provide sufficient theoretical matter to the reader with a satisfactory understanding of the equipment or devices covered. Equipment discussed is of the type that can be purchased by contrast to in-house-designed or constructed apparatus.

The editors express their sincere thanks to the experts who contributed to these collected volumes.

Mahesh V. Bhatia
Paul N. Cheremisinoff

TABLE OF CONTENTS

Page

CHAPTER 1

GAS CONDITIONING

CLARENCE T. MOSS
Niagara Blower Company
New York, NY

The most common use of gas conditioning equipment in chemical plants is to provide close control of both dry bulb and wet bulb temperatures with the result that both dew point and relative humidity are accurately defined. This section will describe equipment available for standard heating, cooling, humidifying and dehumidifying applications without reference to filtration components or cryogenics.

While the fluid to be conditioned will in all cases be considered as air, the principles of air conditioning and heat transfer may be extended to other gases.

APPLICATIONS

Conditioning of air for process and product control may require one or more of the psychometric processes mentioned previously (1). The equipment must be of the proper design and capacity to fulfill these conditions and must take into consideration not only the temperature and humidity in the space but also external loads and internal sources of both heat and vapor.

The relatively simple controls used with ordinary office air conditioning will not be satisfactory in the chemical process applications. Process work generally requires closely held conditions which are vital to the production rate, quality of product and possibly the safety of the operation.

Typical requirements of dry bulb and relative humidity for various products and processes, such as pharmaceuticals and plastics, are given in the 1978 Application Handbook (2), and the Encyclopedia of Chemical Technology (3).

Control of dry bulb temperature only may be provided by various configurations of coils and unit heaters for drying and curing. They may be used for removal of excess moisture in the air or for the prevention of condensation on cold surfaces.

Hygroscopic materials will absorb water from the air (4). Deliquescence, lumping and caking must be prevented. Control of the space conditions is thus important to product quality control, rate of production and packaging requirements.

Removing sensible heat from process gas, including air, and reducing the energy requirements of refrigeration equipment by utilizing a relatively high temperature heat sink, such as well water, are common in the process industries.

Dehumidification has a wide range of applications, from cooling of outside air to below the dew point temperature, so that condensation occurs, to chilling of gas to below-freezing temperatures in photographic film drying. Cellophane, rayon and other man-made fibers, high-speed printing on film, and many individual processes

1

in the pharmaceutical and biological industries require the highly accurate control of temperature and humidity only possible with specialized equipment (1). This is especially important if a chemical reaction is involved which is sensitive to either variable. If the process includes a biochemical reaction or crystal growth, control of both dry bulb and humidity will be considered as they affect rate of reaction and size of crystals. Storage of many chemicals, cleanliness of operating production areas and control of odors and fumes may all necessitate air conditioning.

While the process of absorption is utilized for dehumidification, it is also used in the chemical industries for scrubbing or washing a mixture of gases with a liquid so that one or more of the constituents of the gas will be absorbed into the liquid. Many commercial processes are now available with the aim of removal of pollutants (such as sulfur and ammonia from coke oven gas), by-product recovery and gas separation. The processes, absorbing fluids and equipment for these applications are covered in other literature (5).

The use of adsorbent materials in removal and recovery of solvents has greatly increased due to the enaction and enforcement of environmental protection regulations. These desiccants are also used in dehydration and separation of gases, odor and toxic gas removal (6).

HEATING

Sensible Heating

Equipment available for air-heating purposes varies from individual coils to an assembly of elements such as the Unit Heater, the essential parts being a fan and motor assembly, a heating element and an enclosure.

Coils are of the extended surface type with the secondary surface manufactured from thin metal plates or spiral or ribbon fins. Hot-dip galvanized steel, copper and aluminum are generally used for materials of construction. Process requirements may dictate special materials or finishes such as tinned copper fins, Admiralty or cupro-nickel tubes.

Unit heaters can be classified by one of several methods:

- By heating mediums such as hot water, steam, heat transfer fluid, gas or oil indirect-fired, and electric.
- By type of fan such as propeller, centrifugal and remote air mover.
- By arrangement of the fan and heating medium, either a draw-through or blow-through arrangement.

A great deal of information is available in regard to coil design and selection, heating mediums and variables affecting heat transfer in the 1975 ASHRAE Equipment Handbook (7).

Similar information on unit heaters is presented in the same Volume (8).

Heating with Humidifying

Evaporative air washers may be used to humidify by passing air through finely atomized sprays (spray-type washers) or over wetted cells packed with glass, metal or fiber screens (cell-type washers). The spray-type washers include a high-velocity design where velocities three-to-four-times those of conventional air washers are used to reduce space and weight requirements. Where close control (within $\mp 0.9°$C or $0.5°$F dry bulb temperature and $\mp 0.5\%$ relative humidity) is desired, saturating equipment with a special spray arrangement is available (Figure 1.1).

Figure 1.1 Saturating spray conditioner. (Courtesy Niagara Blower Co.)

Heating with air humidification may be accomplished by preheating the air and passing it through recirculated spray water or by using heated spray water. The former method increases both the dry-bulb and wet-bulb temperatures and lowers the relative humidity but does not alter the humidity ratio (pound of water vapor per pound of dry air). As a result, more water may be absorbed per pound of dry air passing through the apparatus.

By heating the spray water, it is possible to elevate both the dry and wet-bulb temperature above the dry bulb of the incoming air. For example, saturation at 60°C (140°F) is available to meet high-humidity requirements for printing on film such as cellophane. The spray water may be heated by a coil installed in the unit or, if the load is relatively large, it may be heated by a liquid heater in the spray line.

The sprayed coil conditioner shown in Figure 1.2 may be provided with both heating and cooling coils. Utilizing a massive drenching spray for full coil coverage and close contact between air and water, this equipment will give 0.5°C approach between dry-bulb and wet-bulb temperatures and dependable control using simple instrumentation.

Figure 1.2 Drenched coil conditioner. (Courtesy Niagara Blower Company.)

Spray-type washers vary the number of banks of nozzles depending on the degree of saturation desired (Figure 1.3). There is no standardization with this type of equipment. Factors each manufacturer will consider for a specific application include air velocity, spray rate per cubic foot of air per minute, spray pressure and number of stages. Up to 95% saturation efficiency may be obtained.

Figure 1.3 Air washer. (Courtesy Buffalo Forge Co.)

Cell-type washers obtain saturation efficiencies up to 98% by passing the air through cells packed with various types of fill, such as glass, metal or fiber media. Since this equipment does not require atomization of water, sprays may be eliminated in favor of a simple water distribution system over the fill (Figure 1.4).

High-velocity spray-type washers have two types of eliminators, the rotating and fixed-blade arrangements. They also may be designed for increased number of spray banks for greater saturation efficiency.

The most common material of construction is steel with a protective coating of

Figure 1.4 Cell-type washer. (Courtesy Buffalo Forge Co.)

galvanizing or corrosion-resistant finish. All-welded casings, as well as bolted and gasketed panels, are available.

Humidification with air washers is presented in great detail in various handbooks (9, 10).

COOLING

Sensible Cooling

Certain process applications may require only cooling of air. This may be satisfied by simple fan-coil assemblies, with various coolants through the tube bundles or by evaporative cooling.

Coils for sensible cooling of gas may be mounted in the process duct work or purchased as part of a unit comprised of coil, casing and fan section. They may be floor mounted or suspended, blow-through or draw-through designs (Figure 1.5).

Most applications call for an extended surface design with materials of construction similar to heating coils, as well as steel tube-and-fin coils with a hot-dip galvanized surface. Fin spacing, number of rows, gas velocity and type of coolant, as well as coolant temperature, must be considered for each application.

Figure 1.5 Fan cooler. (Courtesy Niagara Blower Co.)

Cooling with Dehumidification

There are several methods of providing dehumidification above the freezing point of water, when defrost is not required, as well as several types of equipment applicable to temperatures below the freezing point, where defrost may or may not be required.

Above Freezing

Air Washers — The same type of equipment used to produce humidification by sprays, with or without integral coils, is applicable to cooling and dehumidifying (11). A decrease in the dry and wet-bulb temperatures will result if the spray water temperature is below the entering air temperature. The air leaving the washer may easily be saturated to within 1.8°C (1°F) of the leaving spray water temperature, giving a performance factor of approximately 0.95, depending upon operating conditions (12). Note that there are two different methods of calculating performance factor, and any specification must clearly state the operating conditions and method of calculation.

Coils — Coils operated so that they cool the gas below the dew point, producing dehumidification, a wetted surface and condensate are still considered as "dry" coils. They may also be sprayed by water or other liquids to aid in dehumidification by absorption or to prevent frost. These are known as "wet" coils.

Dry Operation — Extended surface coils, as used for heating, may also be designed for cooling and dehumidifying. The design parameters cover the same factors such as type and spacing of fins, ratio of fin-to-prime surface area, etc. Chapter 6 of the 1975 ASHRAE Equipment Handbook describes the factors for consideration in selecting a coil, and includes an extensive bibliography (13).

Refrigerants include chilled water, fluorocarbons, ammonia and brine. Volatile refrigerants may operate with a flooded system, dry expansion or liquid recirculation of the refrigerant.

Manufacturers limit the maximum gas velocity across the coil face to eliminate carry-over of water into the duct work or process area, sometimes installing elimi-nators beyond the coil to ensure separation of any droplets from the gas.

Sprayed Coils — Installation of sprays, recirculating pump and drain pan in conjunction with a coil, casing and motor blower results in one of the most ex-tensively-used types of cooling equipment, variously called a spray cooler, spray dehumidifier or spray conditioner (Figure 1.6).

Figure 1.6 Sprayed coil conditioner. (Courtesy Niagara Blower Co.)

Eliminators may be required for certain designs. Coils may be constructed of either prime or finned surface. The latter design is preferred for the large latent loads which this equipment is capable of handling.

Refrigerants are the same as used in dry coils. When cooling coils are used with water sprays, the minimum refrigerant temperature is about $-2°C$ ($28°F$) to pre-vent ice formation on the tube bundle.

Sodium chloride brine may be sprayed over a coil surface to allow the use of a lower refrigerant temperature. This liquid is corrosive and the condensate dilutes the solution, requiring periodic addition of the salt, as well as removal of the weak brine.

Propylene glycol is commonly used as a spray with this type of apparatus to allow low refrigerant temperatures and still control the air leaving condition above freezing. Lithium chloride salt brine is used in a similar manner. Both liquids are used in systems which include a concentration apparatus, with proper controls for maintaining the strength of the liquid by evaporation of water. Low-pressure steam and hot water are generally used in the concentrator as the heating mediums. Solar-heated hot water as low as 66°C (150°F) may also provide economy of operation.

By increasing the concentration of propylene glycol or lithium chloride, water may be absorbed from the gas by the hygroscopic property of the spray liquid and the refrigerating effect of the coil.

Triethylene glycol has a greater hygroscopicity than propylene glycol and is commonly used as an absorbent liquid in spray equipment. Triethylene glycol gives a higher leaving air dry-bulb temperature than propylene glycol for the same leaving dew-point temperature. Concentration equipment is required with this method (Figure 1.7).

Figure 1.7 Laboratory conditioning with absorbent dehumidifier. (Courtesy Niagara Blower Co.)

The glycols are relatively noncorrosive as compared to a salt brine, but cannot reach as low a dew point as a strong brine concentration. The latter system will

9

require close control to prevent salting out of the solution. Both systems are widely used in air conditioning and industrial applications and are discussed further in the section on sorbent dehumidification.

The most common material for casing construction is steel with a corrosion-resistant finish of galvanizing or paint. Highly corrosive brines may use material more resistant to attack such as nickel. Coils may be of aluminum or steel (hot-dip galvanized or coated with a corrosion-resistant finish), depending upon the spray solution and the manufacturer.

Below Freezing

Air Washers — Saturating conditioners of the type shown in Figure 1.1 may be used to saturate air to below freezing temperatures and give precise control of the leaving air condition. This unit, instead of spraying water, will circulate a solution of propylene glycol and water for refrigerants as low as $-23°C$ ($-10°F$) and give saturated air at about $-15°C$ ($5°F$). The moisture absorbed into the antifreeze solution is removed automatically by concentration equipment.

Coils — Dry Operation — As described previously, both finned-surface and prime-surface coils are used in cooling and dehumidifying. The same arrangement of coil, blowers, etc. may be used to give below freezing conditions. However, frost will form on the coil surface and the design must include defrosting capability, as well as additional capacity to maintain design conditions while a portion of the equipment is in the defrost cycle.

Unit capacity is reduced where defrost is required because of the decreased heat transfer coefficient, reduced air volume and greater fin spacing required. An increase in the staging of the defrost cycle will decrease the effect of these factors, but must be considered in the over-all design of the system. Close control of space conditions is difficult with a defrost operation.

Defrosting may be accomplished by outside air, heated air, water spray, electricity or hot gas inside the refrigerant coil. A brine system may use heated brine through the coil for defrosting.

Sprayed Coils — Cooling and dehumidifying with sprayed coils at below freezing temperatures is easily accomplished by the same methods and sprayed-coil equipment described previously for above freezing design. Instead of water, glycol-water and brine solutions are sprayed over either prime-surface or finned-surface tube bundles.

Sodium chloride brine is used above $-12°C$ ($10°F$), and calcium chloride brine down to approximately $-23°C$ ($-10°F$). Lithium chloride brine and glycol-water mixtures (propylene or ethylene) are used in concentrations suffucient to prevent freezing of the liquid at lower refrigerant temperatures. This equipment will easily give $-40°C$ ($-40°F$) leaving gas temperature, removing most moisture and simplifying dry coil operation below this point. A schematic diagram illustrating the components of a glycol system is shown in Figure 1.8 and below-freezing storage

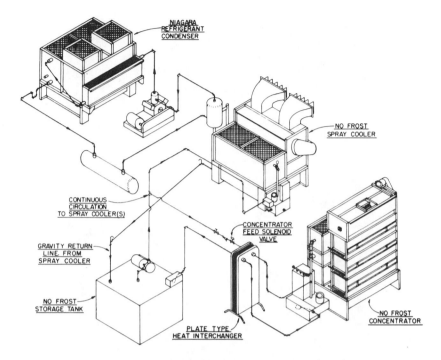

Figure 1.8 Frostless refrigeration with spray cooler and concentrator. (Courtesy Niagara Blower Co.)

equipment, in Figure 1.9. Multiple coolers may be operated from one sump tank and concentrator.

SORBENT DEHUMIDIFICATION

In addition to the drying of gases which may be accomplished in the process of cooling, the use of a sorbent material, either solid or liquid, will fulfill this purpose. There is no direct requirement for cooling unless the heat of sorption, which is always generated in the process, is to be removed. A portion of the heat of interchange between a conditioner and a concentrator in a liquid sorption system is also normally removed.

Liquid sorbents include the solutions of lithium chloride, triethylene glycol, and sulfuric acid, as well as calcium chloride and lithium bromide. These materials change either physically or chemically during the sorption process. Lithium chloride as a solid lies in this category.

Adsorbents do not change physically or chemically during the process. For example, a vapor will be adsorbed into the microscopic pores of activated alumina, silica gel and activated charcoal (6).

Figure 1.9 Warehouse at —23°C with frost-free refrigeration. (Courtesy Niagara Blower Co.)

The selection of the preferred sorbent will depend on the many properties of the multitude of sorbents and the requirements of the specific process application. Among these are:

- Entering and leaving gas temperature and water vapor content
- Coolant temperature available
- Properties of toxicity, flammability, corrosivity and odor
- Viscosity (liquids) and density (solids)
- Availability of regeneration heating medium and its temperature level.

The sorbent capacity of the various materials varies with the type of sorbent, the concentration (with liquids) and the equilibrium temperature of the sorbent. Increasing the latter two variables may result in dehumidifying the gas with little or no depression of the dry-bulb temperature. Figure 1.10 for a triethylene glycol and water solution illustrates this. A spray temperature of 25°C (77°F) will result in the

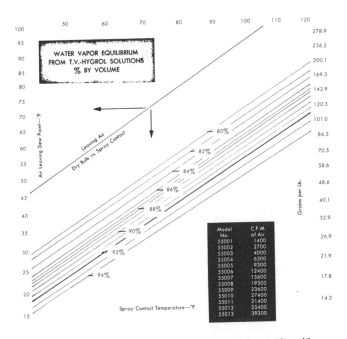

Figure 1.10 Equilibrium curves for absorbent dehumidifier. (Courtesy Niagara Blower Co.)

same leaving air dry bulb (25°C) and a dew point of 5.5°C (42°F), at a concentration of 92%.

Chapter 18 of the 1977 Fundamentals Handbook offers a concise description of these variables (14). Typical performance data is given in the 1975 Equipment Handbook (15).

REFERENCES

1. Kirk, O. and Othmer, *Encyclopedia of Chemical Technology*, Second Edition, Interscience Publishers, div. of John Wiley and Sons, Inc., New York, NY, Vol. I, p. 496.
2. *ASHRAE Handbook and Product Directory, 1978 Applications*, American Society of Heating, Refrigerating and Air-Conditioning Engineers, Inc., New York, NY, 1978, p. 13.2.
3. Kirk, O. and Othmer, *op. cit.*, p. 486.
4. *ASHRAE Handbook and Product Directory, 1978 Applications, op. cit.*, p. 13.4.
5. Kirk, O. and Othmer, *op. cit.*, p. 44.
6. *Ibid.*, p. 460.
7. *ASHRAE Handbook and Product Directory, 1975 Equipment*, American Society of Heating, Refrigerating and Air-Conditioning Engineers, Inc., New York, NY, 1975, p. 9.1.
8. *Ibid.*, p. 27.4.
9. *Ibid.*, p. 4.1.
10. Jorgensen, R. (Ed.), *Fan Engineering*, Buffalo Forge Co., Buffalo, NY, Seventh Edition, 1970, p. 433.

11. *Ibid.*, p. 505.
12. *ASHRAE Handbook and Product Directory, 1975 Equipment, op. cit.*, p. 4.5.
13. *Ibid.*, p. 6.7.
14. *ASHRAE Handbook and Product Directory, 1977 Fundamentals*, American Society of Heating, Refrigerating and Air-Conditioning Engineers, Inc., New York, NY, 1977, p. 18.1.
15. *ASHRAE Handbook and Product Directory, 1975 Equipment, op. cit.*, p. 7.1.

CHAPTER 2

WET SURFACE COOLING

CLARENCE T. MOSS
Niagara Blower Company
New York, NY

INTRODUCTION

Of the types of self-contained, closed-circuit fluid coolers, the dry fintube equipment is described elsewhere in this series. The Wet-Surface fluid cooler also transfers heat to ambient air; however, the tube bundle through which the fluid passes is continuously drenched with water.

This equipment effectively replaces the combination of a cooling tower plus shell-and-tube heat exchanger, generally at a savings in first-cost, operating horsepower and plot area. The Wet-Surface equipment will save at least 95% of the water requirements of a shell-and-tube heat exchanger. It takes far less plot area and operating horsepower than a dry fintube air cooler, when the product temperature closely approaches ambient conditions. For example, the Wet-Surface fluid cooler will easily cool the product to $30°C$ ($86°F$) at $25°C$ ($77°F$) wet bulb temperature. The same approach, using a cooling tower plus shell-and-tube would be very costly. A dry-surface, fintube unit with $35°C$ ($95°F$) air would require several times the air volume and horsepower to even give $40°C$ ($104°F$) leaving product. While each type of heat exchange apparatus has advantages, proper consideration may indicate that Wet-Surface equipment should be selected for cooling of liquids, compressed gases and vapors.

HEAT TRANSFER MECHANISM

Heat is transferred from the fluid inside the tube bundle of a Wet-Surface unit to the ambient air stream by several heat transfer processes. Figure 2.1 illustrates the condensation of a vapor inside a tube.

1. Condensation of vapor on the inner tube surface
2. Conduction through the tube wall
3. Conduction and convection from the outer tube surface through the water film
4. A simultaneous transfer by latent heat, sensible heat and radiation from the wet film to the air stream.

On a clean tube surface, there are three resistances to be considered; namely, that of the condensing film on the inner tube wall, resistance of the metal tube wall, and the resistance of the water film on the outside of the tube. The complex heat transfer relations make it difficult to present Wet-Surface cooler performance in terms of simple parameters such as over-all temperature potential, either dry bulb or wet bulb, or a difference in enthalpy potential. Empirical constants are derived

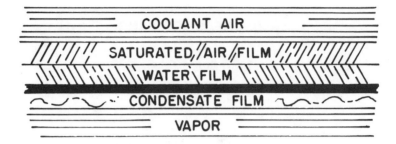

Figure 2.1 Condensation of vapor inside a tube.

from test data. Although theory and analysis methods are useful in interpolating between test points, extension of this data to equipment of other manufacturers is not advisable (1, 2, 3).

RATINGS AND CAPACITY FACTORS

Most process applications do not lend themselves to standardized capacity ratings, curves or tables. They must be calculated for the specific design requirements, with variables such as pressure, temperature, condensing or heat rejection curve, thermal fluid characteristics, design pressure drop, etc.

Factors which affect Wet-Surface cooler performance are operating or ambient wet bulb temperature, air velocity across the coil and, to some extent, both spray loading and elevation or barometric pressure.

A lower wet bulb means a greater driving force or potential between the spray water film and the air, resulting in a lowered spray temperature and increased capacity. The reverse would be noted with high wet bulb temperature, such as the 27°C (80.6°F) value one may expect in the Houston area.

As the wet bulb temperature decreases, each kg (pound) of air can absorb an increasingly smaller number of Joules (Btu) (see Figure 2.2). Though the fans, handling constant volume, are taking a greater weight of air through the unit as the temperature decreases, this increase is far outweighed by the decrease in heat absorbing capability.

The decrease in density of air at high elevations is cancelled by the greater ability to absorb heat than at sea level.

The approach of the leaving fluid temperature to the design ambient wet bulb temperature has a sharp effect on unit capacity. Note Figure 2.3. The effect of the inside coefficient of heat transfer upon over-all heat transfer is shown in Figure 2.4.

An increased spray over that nominally considered satisfactory for complete coverage of the wetted surface affects the capacity favorably to a slight degree. Increased air velocity also has a positive effect, although any manufacturer that requires eliminators will generally be limited by the velocity at which droplets will

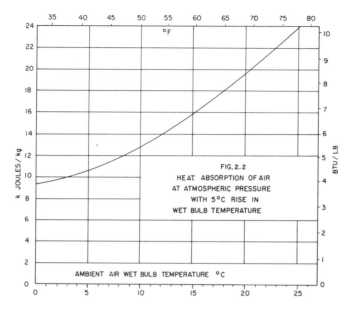

Figure 2.2

PERCENT CAPACITY			
	APPROACH TO WBT		
	15 °C	10°C	5°C
WATER COOLING	100	75	51
COMP. AIR COOLING	100	84	46
WATER VAPOR COND.	100	67	33
AMMONIA COND.	100	69	36

Figure 2.3 Effect on capacity of approach of
product temperature to ambient wet bulb
temperature.

not carry over. This is the case with equipment which has air up through the tube
bundle and eliminators. The upward circulation of air against the spray tends to
sustain the film of the water over the tube surface, as in contrast to air passing
down with the spray. Here, the flow of water over the coil is accelerated, resulting
in a somewhat more efficient apparatus.

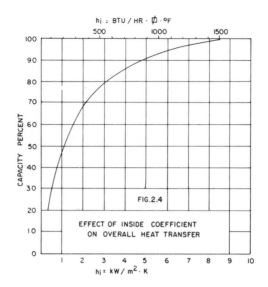

Figure 2.4

CONSTRUCTION

All Wet-Surface exchangers have four basic components:

1. A pan from which water is circulated
2. A pump and recirculation system to thoroughly wet
3. A tube bundle into which the fluid to be cooled is directed and over which is passed a quantity of ambient air propelled by
4. A fan assembly

A small portion of the spray water is evaporated from the film which floods the surface of the tube bundle, rejecting approximately 2442 kJ/kg (1050 Btu/lb) of water evaporated, cooling the spray water and tube surface and removing heat from the fluid inside the tube.

Wet-Surface cooling equipment is available with air passing up through the coil, against the spray, and also with air being drawn down through the coil and spray. Fans may be forced-draft or induced-draft, centrifugal, axial flow or propeller, while sprays always flood the coil from above.

Tube Bundles

Design of the heat transfer surface varies with the product to be cooled and/or condensed. Variation in construction for cooling of liquids, gases and vapors will be shown in more detail in later sections.

Manufacturers use varying tube diameters and wall thickness, spacing and pitch, depth of tube bundles, number and direction of fluid passes and mass velocity of air over the bundle. Tube bundles are of three basic designs; namely, serpentine with welded headers, serpentine with removable cover plate for the header, and straight-through design with tubes rolled into tube sheets with cover plates removable over the entire face of the bundle. Wet-Surface tube bundles have a high external coefficient, compared to a similar tube operated dry. Therefore, there is no need for external surface such as fins. However, equipment may be designed for dry operation at reduced loads and low ambient dry bulb, which can utilize a finned-surface arrangement. This operation saves spray pump horsepower, make-up water and the necessity for freezing protection of the liquid in the pump basin, while eliminating the water vapor plume which is evident from evaporative equipment during low temperatures. These bundles are available in all-copper or all-steel, hot-dip galvanized after fabrication.

Tube bundles of the above design may not fall under the requirements of the ASME Code, Section VIII; however, design and stamp are generally available as options.

Inside fouling factors are specified according to process fluid requirements (4). An outside fouling factor of 0.0005 is generally satisfactory if there is proper blowdown and water treatment.

Tube Wetting

A more appropriate term would be "Tube Drenching," in that properly flooding the tube surfaces with water is essential to provide rated capacity. The latter is affected by a reduced coefficient for dry areas, as well as the steady increase of scale on the outside of the tube (with lowered heat transfer capability) as water evaporates from the warm tube surface.

The two basic means of distribution of water over the tube bundle are spray nozzles and troughs, the latter having a saw-tooth edge which distributes rivulets of water at varied intervals along the pipe surface. The manufacturers utilizing this method indicate that the upward draft of circulated air breaks up the water streams, creating a turbulent flow of spray water over the tube surfaces.

Equipment manufacturers utilizing nozzles generally resort to such materials as bronze, brass or plastic of varying orifice diameter and spray pattern. Manufacturers' literature for equipment used with commercial applications indicates a spray rate of 0.122 m^3 per min per sq meter of face area of their bundles (3 gpm per sq ft). It is recommended that this spray rate be increased for all applications, especially for industrial and process work. This is vital in sustaining design performance from evaporative equipment.

In the same general category as spray systems would be that of the strainers which attempt to keep the distribution systems clear of bugs, leaves, etc. The strainer should be placed such that it may be cleaned without interruption of the operation, capable of being handled by one individual, and of sufficient capacity that it will require an ordinary cleaning schedule.

Fan Arrangements

Centrifugal, axial flow or propeller-type fan arrangements are utilized whether the air is induced-draft or forced-draft. Materials for the former are normally steel, either hot-dip galvanized or with a painted finish, or of stainless steel. Propeller and axial fan arrangements can be furnished of cast aluminum (with or without a corrosion-resistant finish), stainless steel or reinforced plastic. Centrifugal arrangements use one or two fans per motor, or an extended shaft, normally constructed of large-diameter hollow tubing for rigidity, with one fan motor per side. Centrifugal and axial flow fans may be direct-connected or driven by belts and pulley arrangements. Maintenance is normally in proportion to the number of fan bearings, motors and V-belts, and center bearings should be avoided.

Until a few years ago, the noise level was only of nominal importance. However, with the promulgation of the OSHA regulations, and increasing incidence of directives to users concerning sound levels, consideration must be given to proper compliance. Centrifugal fans have a low sound level and generally a higher brake horsepower draw than propeller-type or axial flow fans, which will have a lower power requirement but higher sound level. With centrifugal fans, it is not difficult to be well within the OSHA regulations of 90 dBA for an exposure of eight hours. The use of axial flow fans, except when the unit is situated outside away from an occupied site, may require a special low-speed selection to meet sound level requirements.

On large equipment and industrial applications where there is a multiplicity of fans on one unit, an alternative is offered for one large-diameter, gear driven, slow-speed, multiblade propeller-type arrangement which minimizes horsepower, as well as installation cost (see Figures 2.5, 2.6, 2.7).

Eliminators

The configuration which utilizes air up through the coil and sprays, by the very nature of the case, requires eliminators which may be manufactured of steel with various finishes, stainless steel or plastic. They are designed to divert the leaving air away from the air intake so as to reduce recirculation of moist air into the machine. The normal air velocity leaving an eliminator does not exceed 200 m/min (650 ft/min). As a comparison, equipment which does not require eliminators (the downdraft design) can have an exit velocity of approximately 458 m/min (1500 ft/min) which reduces recirculation.

Any eliminator will have an inherent resistance. The more effective the eliminator, the greater the resistance under normal consideration, requiring additional horsepower to the fans.

The amount of air circulated through the equipment is limited by the maximum velocity through the eliminators. The unit capacity may also be limited by this same restriction, since more air at a higher velocity will carry a greater number of Joules (Btu) for a specific amount of tube surface.

Figure 2.5 Shows multiple fans on one unit installation. (Courtesy of Niagara Blower Co.)

Figure 2.6 One large-diameter, slow speed fan arrangement. (Courtesy of Niagara Blower Co.)

Figure 2.7 Another example of 3-large diameter fans. (Courtesy of Niagara Blower Co.)

Casing

Steel enclosure or casing panels are used, normally hot-dip galvanized after fabrication or with various painted finishes. Fasteners may range from hot-dip galvanized bolts, nuts and washers to cadmium-plated or stainless-steel fasteners, to metal screws of various materials. Panels may be fastened such that they may be individually removed for cleaning of interior surfaces and repainting, without disturbing adjacent sections. Stainless steel or aluminum are also offered where service requirements specify.

Access doors in the casing must be large enough to provide interior accessibility, rather than just serve as inspection doors. It may be necessary to enter the pan of the unit, to reach the sprays, fans and fan motors. Regardless of the construction, such accessibility is paramount.

Spray Pumps and Basin

Low-head, high-volume spray pumps are used, either the vertical-submerged or the floor-mounted, direct-connected type. Basins may be an integral part of the unit design, supporting the tube bundle and casing. As an alternate, equipment may be supported on structural steel, with a drain basin of various materials suspended below the equipment. Spray water collected here passes to a common sump at ground level and may be sized for one or more cells. Concrete basins are commonly used.

CAPACITY MODULATION

The product temperature leaving the tube bundle normally decreases as either the load or the ambient wet bulb temperature decreases. If this is undesirable, a method of controlling the capacity must be available. An example of this is found in the refrigeration industry, where low refrigerant condensing pressures must be avoided to assure proper operation of expansion valves or other refrigerant control devices. Here, a condensing pressure-control switch is used to reduce capacity in one of several ways:

1. Intermittent operation of the fan
2. A damper system of one of several types
   - Two-position dampers for throttling air flow
   - Modulating dampers for increased range of air volume control
   - A recirculating system of dampers, maintaining air flow and modulating the wet bulb of the air passing over the tube bundle to maintain a constant condensing pressure
3. Multi-speed fan motors

Reduction of air volume on equipment utilizing water distribution troughs would adversely affect water distribution over the tube bundle and likely result in partial wetting of the surface, with attendant negative effects on capacity and increased scaling.

It is noteworthy that standard references twenty years ago recommended intermittent operation of the water circulating pump, or both the pump and the blowers, for capacity reduction (5). Most manufacturers, as well as the 1975 ASHRAE Handbook, recommend that this method not be utilized (6). Each time the pump is stopped and the tube surface dries, a thin film of scale may be deposited with a continual build-up over a period of time, reducing both heat transfer and air volume through the unit. The condenser, operating dry, has a greatly reduced capacity; the head pressure will quickly increase, resulting in short cycling of the pump system, rapid build-up of scale deposits and a shortened life for the circulating pump assembly and starter.

The dry portion of such a cycle may be lengthened considerably by using a finned-surface condensing coil. However, scaling is still certain to occur with intermittent operation of the spray pump. If a reduced load is expected in the winter, the unit can be designed with a finned-surface coil, so that it may be operated completely dry for several months in winter climates.

When multi-fan arrangements are used, one or more may be operated intermittently for reduced capacity, with proper baffling of the air so as to prevent short cycling through the inoperative section.

FREEZE PROTECTION

Wet-Surface heat exchangers are generally located outdoors, where they may be susceptible to freezing and may be protected by one of the following methods.

1. With a dry-pan operation, the circulating pump and sump tank are placed below the equipment in a heated area, into which the spray water falls by gravity. Heaters may be placed in the sump, if necessary.
2. Either electric heaters or a steam coil may be placed in the spray pan of the unit, controlled by a pan thermostat to prevent freezing of the pan water when there is no load. A steam injector may also be used to heat the pan water directly.
3. The full damper system, mentioned previously under Capacity Control, automatically results in the entire unit being enclosed (see Figure 2.8).

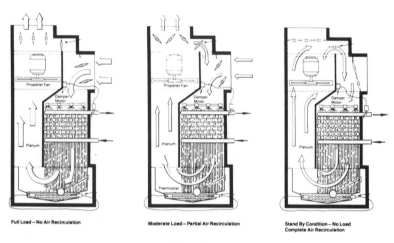

| Full Load — No Air Recirculation | Moderate Load — Partial Air Recirculation | Stand By Condition — No Load Complete Air Recirculation |

Figure 2.8

Intake and discharge dampers, under idle position, prohibit air from blowing into the unit through either opening. This minimizes heat loss and will allow operation of warm spray water over the tube bundle. The spray water is maintained at $5°C$ $(41°F)$ by pan heaters and thermostats. Safety thermostats and back-up controls are normally specified when operation in below-freezing climates is expected.

COOLING OF LIQUIDS

Applications

The Wet-Surface Aircooler for removing heat from a liquid has a myriad of uses in the chemical industry. Wherever heat must be rejected in a process at a low temperature level (approaching ambient air conditions), this method should be investigated.

While no list will be complete, the following examples of liquid cooling may serve to indicate the range of uses to which this equipment may be applied.

M.E.A. solutions
Methanol, Ammonia liquor
Deionized, demineralized water
Heat transfer fluids, autoclaves
Liquid fertilizer
Polyalkylene Glycol
Distillates
Propane, butane
Lean oil
Water for reactor vessels, barometric condensers, blending tanks, wash towers

Tube Bundle Design

The simplest tube bundle design for cooling liquids is the serpentine arrangement, where each circuit is welded into a common header as shown in Figure 2.9. These tube bundles are generally used with clean fluids such as water and glycols. While normal design allows for 1379 k Pa (200 psi) operating, 2068 k Pa (300 psi) test, the type of header and tube wall may be varied to allow for operating pressures of 34,474 k Pa (5,000 psig). These bundles may be chemically cleaned in place if necessary. While the most common material of construction is steel, with a finish of hot-dip galvanizing, tubes and headers may be manufactured with any materials which may be satisfactorily bonded.

Figure 2.9 Tube bundle design for cooling liquids in serpentine arrangement. (Courtesy of Niagara Blower Co.)

A variation of the above allows for flushing individual circuits by the use of removable cover plates for one or both headers (see Figure 2.10). These bundles are

Figure 2.10 Removable cover plates. (Courtesy of Niagara Blower Co.)

normally designed for approximately 1034 k Pa (150 psig) operating pressure.

Where there is a tendency to foul the inside tube surface, the bundles may be constructed of straight tubes rolled into plate tube sheets with removable cover plates (Figure 2.11). The latter is a flat-plate design for liquid cooling. Most applications do not exceed 1206 k Pa (175 psig) operating pressure. Removable cover plates allow mechanical cleaning of tube interiors plus rerolling and replacement of tubes, in place, if necessary.

Materials of construction are dependent upon process requirements. For ordinary water cooling applications, steel tubes or pipes, hot-dip galvanized on the inside and outside, are rolled into galvanized tube sheets which have been hot-dip galvanized prior to tube-hole drilling and grooving. The cover plates may be galvanized but may have any of several available finishes or coatings. The entire tube bundle may be manufactured of stainless steel, admiralty, Monel, etc.

Figure 2.11 Tube bundles of straight tubes rolled into plate tube sheets. (Courtesy of Niagara Blower Co.)

COOLING OF COMPRESSED GASES

Applications

Air compressed to about 860 k Pa (125 psig) for general plant use is the most common application in cooling of compressed gases. The primary heat load is the sensible cooling of the gas, although latent heat may be rejected as a condensible vapor such as water is cooled. Fluids, such as hexane and natural gas, are representative of other gas-cooling applications.

Tube Bundle Design

Figure 2.12 shows equipment for cooling both compressed air and water for intercoolers plus jackets, and illustrating the construction of the tube bundle and headers to allow sufficient flow area for an economical pressure drop. Correct header design will result in the proper balance between heat transfer and friction loss, with a normal pressure drop of approximately 7 k Pa (1 psi).

The compressed air tube bundles are generally constructed of pipe, hot-dip galvanized inside and outside, which gives protection against condensate corrosion on the interior surfaces. The pipes are rolled into hot-dip galvanized tube sheets, and cover plates are removable.

Water cooling coils may be of serpentine design if the cooling circuit is completely closed. Removable cover plates, with pipes or tubes galvanized inside as well as outside, are strongly recommended where a circuit open to the atmosphere allows oxygen to enter the cooling water.

27

Figure 2.12 Equipment for both compressed air and water cooling.
(Courtesy of Niagara Blower Co.)

COOLING AND CONDENSING OF VAPORS

Applications

The following examples are representative of applications for condensing vapors in the chemical industry:

Carbon Dioxide saturated with water vapor
Turbine exhaust steam
Isobutane, propane, light hydrocarbons
Propylene, natural gas
Ammonia Synthesis gas
Overhead condenser for soda ash crystallizer, aluminum sulfate crystallizer, polymerization reactor, methanol stripper
Hexane, heptane, amylene
Methyl chloride, alcohols

Tube Bundle Design

Vapors at relatively low pressures, such as ammonia at 1275 k Pa (185 psig) and fluorocarbon refrigerants, are cooled and condensed in serpentine coils as described in the section for cooling of liquids. Ammonia condensers with de-superheating coils and oil-out traps are shown in Figure 2.13).

Figure 2.13 Ammonia condensers with desuper-heating coils and oil-out traps. (Courtesy of Niagara Blower Co.)

A process which requires cooling a vapor at relatively high pressure, such as 34,475 k Pa (5,000 psig), will require special design and materials. Such a tube bundle, used in ammonia synthesis gas processing is shown in Figure 2.14.

Vacuum operation requires a design with a great number of tubes in parallel for low pressure drop (see Figure 2.15). In a partial condenser application, the tube bundle will normally be one pass. A vacuum application for condensing steam under high vacuum may have two passes, the second of which is a subcooling section for noncondensibles. The temperature of the mixture of water vapor and noncondensibles being drawn out by the vacuum apparatus is reduced well below the dew point equivalent to the operating pressure. This minimizes the size and energy requirements for the equipment maintaining the vacuum, such as a steam ejector. Figure 2.16 illustrates this condensing and subcooling arrangement.

While most such tube bundles are constructed of steel with hot-dip galvanized

Figure 2.14 Ammonia synthesis gas processing tube bundle. (Courtesy of Niagara Blower Co.)

finish, process requirements many times specify one of the stainless steel alloys. Admiralty and cupro-nickel are two other common selections.

INSTALLATION, OPERATION AND MAINTENANCE

Location

While most Wet-Surface equipment will be installed outside, the advantage of having a specific unit for individual processes may allow the equipment to be located indoors. The discharge air must be carried to the outside and installed such that moist air will not be recirculated to the air intake. Warm, moist air may condense in the ducts if these pass through a cool area. Drainage provisions must be considered.

Figure 2.15 Vacuum operation design. (Courtesy of Niagara Blower Co.)

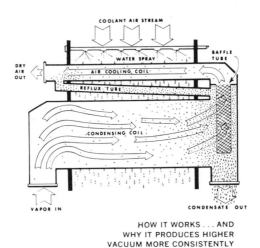

HOW IT WORKS . . . AND
WHY IT PRODUCES HIGHER
VACUUM MORE CONSISTENTLY
THAN WATER COOLED CONDENSERS

Figure 2.16 Condensing and subcooling arrangement.

If the unit is designed for dry operation in the winter, this warm air can be returned to the building. It is generally not advisable to return warm, moist air from a Wetted-Surface operation to an occupied space. Ductwork is extensive, and inside location may require centrifugal fans which can be rated to handle the extra resistance.

Outside installations should consider the relationship of the prevailing wind to adjacent structures, such as buildings, which might cause the warm, moist discharge air to recirculate to the intake. If the unit is not designed for proper operation during freezing temperatures, alternative provisions, such as draining, will be necessary. This normally allows a corrosive action to start on the interior tube surfaces each time such drainage occurs.

Regardless of the installation, the intake air should not pick up corrosive contaminants, such as smokestack emissions, sulfur fumes, etc., which can make short work of galvanized condensing coils.

With multiple installations, align the equipment so that the prevailing wind may least adversely affect the wet bulb of the ambient air on the lee unit.

Piping

Wet-Surface Condensers must be piped such that the condensate will continually drain. Air and other noncondensibles must be purged from the tube bundle. Standard piping practices are illustrated in the 1976 ASHRAE Systems Handbook, Chapter 27 (7). Recommendations on individual trapped drop legs for each tube bundle on multiple condensers, as well as recommendations on equalizer line sizing between the receiver and the hot-gas line of the condenser, are also given in the same reference.

A purge unit is recommended for installation in the system to remove noncondensible gases that would cause operation at higher than normal condensing pressures. Noncondensibles are accumulated in the discharge header of Wet-Surface Condensers. It is standard practice to have a purge connection off the top of this header for connection by the customer. The presence of noncondensible gases in the condenser not only causes higher head pressures, but increases the work of compression and operating horsepower, reduces capacity and promotes oxidation of the lubricating oil. It is one of the most common causes for apparent lack of capacity in this type of equipment.

Piping to the connections of equipment with removable cover plates must be handled in such a way that there is clearance to remove the cover plate after disconnecting, as well as insuring freedom of access to the tubes themselves. Figure 2.17 illustrates one such method.

Blowdown and Water Treatment

Since normal make-up water has impurities which, if not controlled, will eventually cause scale formation, corrosion, reduction in heat transfer and air volume, a positive bleed is necessary. It is possible to remove the impurities by draining out all of the water in the sump or spray pump at regular intervals. However, some form of

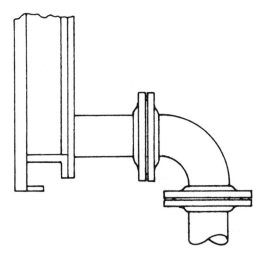

Figure 2.17 Piping with interconnections.

automatic bleed is recommended. The amount of water wasted will depend upon the concentration of impurities and other chemical properties of the make-up water supply. It can be as little as one-half the amount evaporated or triple this amount. Regardless of this, it is still only 5% or so of the water consumed by a once-through water-cooled condenser.

While the hardness expressed in parts/million of total hardness of tap water for representative cities in the United States is given in Table 1, Chapter 16 of the 1975 ASHRAE Equipment Handbook, it is normally recommended that a customer consult a local water treatment firm (6). Automatic treatment equipment is available which gives excellent results, once the proper adjustments are made to the controls. In this case, the blowdown rate is also controlled automatically by a timer and solenoid valve. Biocides may also be included as part of the treatment process. A qualified representative of the treatment firm is the best asset to maintaining proper condition of the coils.

Maintenance

Each manufacturer has recommendations concerning proper maintenance. It is extremely important that a maintenance program be established, so that regular inspection and maintenance will occur. A summarization of items which should be checked periodically would be:

- Lubrication of fan and motor bearings, adjustment of fan belts
- Check fan blades for build-up of scale, possibly evidenced in vibration
- Contaminated circulating water in the sump, fouled strainers, clogged nozzles

- Inadequate treatment and/or blowdown, as evidenced by build-up of scale on the coil surfaces
- Proper water level and make-up valve operation in the sump

Yearly maintenance includes inspection of casing and pan finish, with prompt repairs where needed.

REFERENCES

1. Goodman, W., "Air Conditioning Analysis" (New York, Macmillan Co.) (1947).
2. Kals, W., Wet-Surface Aircoolers, *Chemical Engineering* (New York, McGraw-Hill, Inc.) (1971).
3. Kals, W., Wet-Surface Aircoolers: Characteristics and Usefulness, *Report No. 72-HT-28*, The American Society of Mechanical Engineers, New York (1972).
4. Peters, M. S. and Timmerhaus, K. D., *Plant Design and Economics for Plant Engineers*, McGraw-Hill, New York (1969).
5. *The Refrigerating Data Book*, Basic Volume, Seventh Edition, The American Society of Refrigerating Engineers, NY (1951).
6. *ASHRAE Handbook and Product Directory, 1975 Equipment*, American Society of Heating, Refrigerating and Air Conditioning Engineers, Inc., NY (1975).
7. *ASHRAE Handbook and Product Directory, 1976 Systems*, American Society of Heating, Refrigerating and Air Conditioning Engineers, Inc., NY (1976).

CHAPTER 3

AIR COOLED HEAT EXCHANGERS

D. F. FIJAS
American Standard Inc.
Heat Transfer Division
Buffalo, NY

INTRODUCTION

Air cooled heat exchangers are used to transfer heat from a process fluid to ambient air. The process fluid is contained within heat conducting tubes. Atmospheric air, the coolant, is caused to flow perpendicularly across the tubes in order to remove heat. In a typical air cooled heat exchanger, the ambient air is either forced or induced by a fan or fans to flow vertically across a horizontal section of tubes. For condensing applications, the bundle may be sloped or vertical. Similarly, for relatively small air cooled heat exchangers, the air flow may be horizontal across vertical tube bundles.

In order to improve the heat transfer characteristics of air cooled exchangers, the tubes are provided with external fins. These fins can result in a substantial increase in heat transfer surface. Thus, the design of finned tubes is highly important in air cooled heat exchangers and will be discussed in detail. Parameters such as bundle length, width and number of tube rows vary with the particular application as well as the particular finned tube design.

The first applications of air cooled heat exchangers were in areas where the supply of cooling water was scarce or non-existent. This condition more or less demands the use of ambient air as the coolant fluid. However, in recent years, air cooled heat exchangers are being used in all areas as a viable alternative to water cooled heat exchangers (1, 2, 3, 4, 5, 6).

The choice of whether air cooled exchangers should be used is essentially a question of economics (7, 8) including first costs or capital costs, operating and maintenance expenses, space requirements, and environmental considerations; and involves a decision weighing the advantages and disadvantages of cooling with air.

ADVANTAGES OF AIR COOLED HEAT EXCHANGERS

The advantages of cooling with air may be seen by comparing air cooling with the alternative of cooling with water. Many of these points will be discussed later for individual air cooled heat exchanger components, but are listed here in summary.

1. Since water is not used as the cooling medium, the disadvantages of using water are eliminated:
   - Eliminates high cost of water including expense of treating water;

- Thermal or chemical pollution of water resources is avoided;
- Installation is simplified due to elimination of coolant water piping;
- Location of the air cooled heat exchangers is independent of water supply location;
- Maintenance may be reduced due to elimination of water fouling characteristics which could require frequent cleaning of water cooled heat exchangers.

2. Air cooled heat exchangers will continue to operate (but at reduced capacity) due to radiation and natural convection air circulation should a power failure occur.
3. Temperature control of the process fluid may be accomplished easily through the use of shutters, variable pitch fan blades, variable speed drives, or, in multiple fan installations, by shutting off fans as required.

DISADVANTAGES OF AIR COOLED HEAT EXCHANGERS

The disadvantages of air cooled heat exchangers should also be considered in chemical equipment design:

1. Since air has relatively poor thermal transport properties when compared to water, the air cooled heat exchanger could have considerably more heat transfer surface area. A large space requirement may result.
2. Approach temperature differences between the outlet process fluid temperature and the ambient air temperature are generally in the range of 10 to 15°K. Normally, water cooled heat exchangers can be designed for closer approaches of 3 to 5°K. Of course, closer approaches for air cooled heat exchangers can be designed, but generally these are not justified on an economic basis.
3. Outdoor operation in cold winter environments may require special consideration to prevent freezing of the tube side fluid or formation of ice on the outside surface.
4. The movement of large volumes of cooling air is accomplished by the rotation of large diameter fan blades rotating at high speeds. As a result, noise due to air turbulence and high fan tip speed is generated.

MAJOR COMPONENTS OF AIR COOLED HEAT EXCHANGERS

The major components of air cooled heat exchangers include the finned tube, the tube bundle, the fan and drive assembly, an air plenum chamber, and the overall structural assembly. Each of these items will be discussed below.

Finned Tubes

Common to all air cooled heat exchangers is the tube through which the process fluid flows. To compensate for the poor heat transfer properties of air which flows across the outside of the tube and to reduce the overall dimensions of the heat exchanger, external fins are added to the outside of the tube.

A wide variety of finned tube types (see Figure 3.1) are available for use in air cooled exchangers (9). These vary in geometry, materials, and methods of construction which affect both air side thermal performance and air side pressure drop. In addition, particular combinations of materials and/or fin bonding methods may determine maximum design temperature limitations for the tube and limit environments in which the tube might be used. The use of a particular fin tube is essen-

a.   Metallurgically bonded Fins

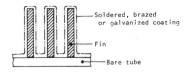

Soldered, brazed
or galvanized coating

Fin

Bare tube

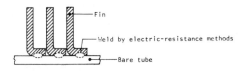

Fin

Weld by electric-resistance methods

Bare tube

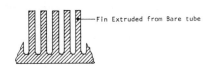

Fin Extruded from Bare tube

b.   Mechanically Bonded Fins
     Imbedded Fins

Fin

Fin is inserted in groove machined into bare tube, after which tube
metal is displaced to hold fin in place.

Bare tube

Tension Wrapped Fins

Fin

"L-foot" fin is wrapped under tension to make bond between fin and
tube via contact pressure.

Bare tube

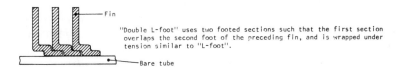

Fin

"Double L-foot" uses two footed sections such that the first section
overlaps the second foot of the preceding fin, and is wrapped under
tension similar to "L-foot".

Bare tube

Plate type fins - Normally bonded to the tube by expanding the tube against the fin, the plate type
fin similar in appearance to the "L-foot" above. Fins are individually formed prior
to insertion over the tube.

Figure 3.1 Finned tube types.

tially a matter of agreement between the air cooled heat exchanger manufacturer and the user.

Finned tubes may differ in the means by which the fins themselves are attached or bonded to the bare tube. This bond may be mechanical or metallurgical in nature. Metallurgical bonds are those in which a solder, braze, or galvanizing alloy coats the fin and bare tube or in which the fin is welded to the tube. Fins which are extruded or machined from the base tube and are therefore integral with the tube may also be considered as having a metallurgical type bond.

Mechanically bonded tubes may be of two types. First, "imbedded" or "grooved" tubes are formed by machining a helical groove along the length of the tube. The fin is located in the grooved and wrapped around the tube after which the tube material is deformed at the base of the fin. This procedure holds the fin in place and in contact with the tube. Since a groove approximately 0.025 to 0.030 cm deep is cut in the tube, the wall thickness of this type tube must be heavier by that amount.

Secondly, mechanically bonded tubes may be obtained by mechanically stressing the fin material and/or the tube material to hold the two elements in pressure contact with one another. So called tension wound fins are formed by winding the fin material under tension in a helical manner along the length of the tube. This method stresses the fin material to maintain contact with the tube. The ends of the fins must be held in place to keep the fins from loosening. This may be done by means of stapling, brazing, soldering, welding or any other way to keep the fins from unwrapping.

Individual fins may be preformed and inserted over the tube after which the mechanical bond may be obtained by either shrink fitting the fins onto the tube or by expanding the tube radially outward to make pressure contact with the fin material. The means to expand the tube may be hydraulic by pressurizing the tube beyond its yield point; or it may be of a mechanical nature in which an oversized ball or rod is pushed through the length of the tube forcing the tube material outward against the fin.

Tubes whose fins are integral with the tube may also be classified as a mechanical bond type if a liner tube is used inside the finned tube. A liner tube of another material may be used for compatibility with the tube side process fluid. The contact between the two materials could be formed by expanding the liner tube or by drawing the outer finned tube down over the liner.

The selection of a particular bonding method and/or fin geometry may depend upon the process conditions to be met and upon the environment to which the tubes will be exposed.

The operating temperatures of the exchanger, including upset or transient conditions may affect the bonding method which can be used for the finned tubes (10). In order to maintain design thermal performance, the bond between the fin and the tube must not deteriorate due to a loosening of the fin which could result from unequal thermal expansion of the fin and tube materials.

In order to avoid this degradation of tube performance, mechanically bonded tubes of the tension type are normally limited to temperatures of 400 to 600°K; and mechanically bonded grooved fin types from 600°K to 700°K.

Metallurgically bonded tubes are limited to temperatures below the melting point of the bonding alloy or to a temperature dependent upon the physical properties of the tube and fin materials.

The operating environment may influence the choice of materials used and the shape of the fin. Aluminum is very often satisfactory as a fin material, although copper, steel and stainless steel fins are also used. The fin shape may be of edge-type, L-foot type or double L-foot design. The edge type is used for the grooved fin tube and in cases where the base tube is not subject to corrosion. The L-foot fin covers the tube more or less completely to protect the base tube against corrosive attack, but still leaves a potential corrosive site at the base of the fin adjacent to the preceding fin. The double L-foot is intended to provide complete coverage of the tube where corrosion would otherwise be a problem. Where corrosion is troublesome, soldered or galvanized tubes may offer a solution.

Having selected the type of fin construction, materials, and bonding techniques, the determination of tube diameter, fin thickness, fin height and fin spacing must be performed for the particular process conditions under consideration.

Dimensions of finned tubes currently used in industrial applications are results of past experience in the design of air cooled heat exchangers. Tube diameters range from about 1.905 cm (0.75 in.) to 5.08 cm (2.0 in.). Helically wrapped fins are fabricated such that the fin height can be between about 3/8 to 3/4 of the tube diameter, but limited because of fabrication requirements to a maximum of about 2.54 cm (1.0 in.) in height. Fin spacings vary between about 275 and 450 fins per meter of tube length, while fin thicknesses range from 0.025 to 0.075 cm. For particular cases these parameters may be varied further.

Surface area ratios of total surface obtained by adding external fins compared to bare tube surface range from about 7 to 40 depending upon the particular dimensions used. The surface area density for helically wound tube can be calculated by:

$$A_{fin}^1 = \frac{\pi}{2} N_f (d_f^2 - d_r^2 + 2d_f t_e)/P_t/100 \qquad (1)$$

$$A_r^1 = \pi d_r (1 - N_f t_r)/P_t/100 \qquad (2)$$

and

$$A_o^1 = A_{fin}^1 + A_r^1 \qquad (3)$$

where the total heat transfer surface per square meter face area per row of tubes, A_o^1, is the sum of secondary fin surface, A_{fin}^1, plus prime tube surface, A_r^1. Then, the total air cooled exchanger outside surface is:

$$A_o = A_o^1 \times F.A. \times N_r \qquad (4)$$

Figure 3.2 shows the surface area increase due to the addition of external fins for two typical tube diameters.

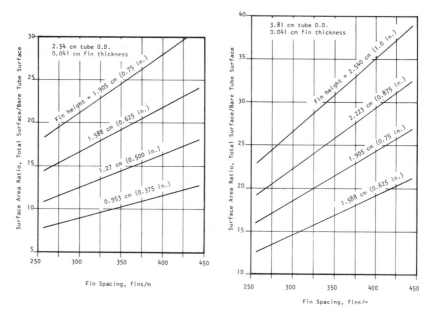

Figure 3.2 Surface area ratios for typical helically wound plain fins.

For L-foot fins, the root diameter of the prime tube should be considered as the tube diameter plus twice the thickness of the foot. Similarly, the double L-foot root diameter should be increased by four times the foot thickness.

Similar expressions for surface area can be determined for other types of fin tubes including square or rectangular fins, segmented fins, and oval shapes tubes and fins.

Tube Bundle

The finned tubes are assembled into the tube bundle. Tube lengths range from about 1.83 m long to as much as 12.2 m long. The number of tube rows deep in the bundle is a function of the performance required and generally ranges between 3 and 30.

Typical bundle configurations (11) are illustrated in Figures 3.3 and 3.4. The ends of the tubes are not finned. This permits the tubes ends to be inserted into tubesheets located at each end of the bundle. The tubesheets separate the cooling air on the fin side from the process fluid on the tube side. Generally, the tube ends are roller expanded into the tube holes in the tubesheet to form the joint, although for higher pressure applications these may be welded joints.

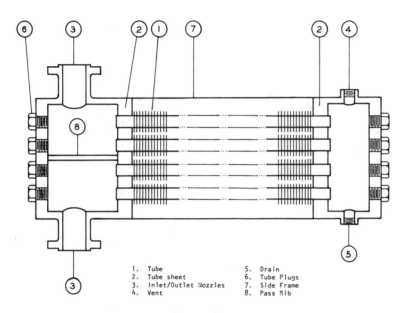

1. Tube	5. Drain
2. Tube sheet	6. Tube Plugs
3. Inlet/Outlet Nozzles	7. Side Frame
4. Vent	8. Pass Rib

Figure 3.3 Typical tube bundle (two pass) using box headers with tube plugs opposite each tube end.

The tubesheets are attached to tube side headers which contain the tube side fluid and distribute it to the tubes. The headers may be designed to permit any number of tube side passes for the process fluid. For multipass tube bundles, the headers contain partition plates which divide the bundle into separate passes. However, these may be limited by the operating temperature conditions. If there is a large temperature difference per pass, then the hotter tubes may expand lengthwise to a much greater extent than the tubes in succeeding passes. This could result in high stresses on the tube joint resulting in leakage at the joint. Tube expansion may be calculated by:

$$\Delta L = \alpha \, L \, \Delta t_{amb} \tag{5}$$

If differential expansion between passes is excessive, split headers may be necessary. The tube bundle is normally permitted to float independently of the supporting

1. Tube
2. Tubesheet
3. Inlet/Outlet Nozzles
4. Pass Rib

5. Gasket
6. Removable Cover
7. Removable Bonnet
8. Side Frame

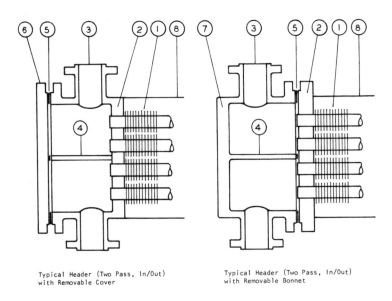

Typical Header (Two Pass, In/Out)
with Removable Cover

Typical Header (Two Pass, In/Out)
with Removable Bonnet

Figure 3.4 Alternate header construction types.

structure due to overall bundle expansion. Values of the thermal expansion coefficient, α, for some common materials are shown on Table 3.1.

Table 3.1. Mean Coefficients of Thermal Expansion (12).

Material	α, m per m per °K x 10^6 Expansion Coefficient between 294°K and temp.:			
	400°K	500°K	600°K	800°K
Carbon Steel	11.7	12.3	13.0	14.3
Austenitic Stainless Steel	16.8	17.5	17.8	18.5
Aluminum	23.7	24.6	25.6	
Copper	17.5	17.8	18.1	

End plates on the tube side headers frequently include removable plugs. These can be pipe-tap plugs or straight threads with gasket seals. The plugs are located opposite each tube end to permit access to each tube for rerolling of the tube to tubesheet joint should leaks occur and for cleaning of the tubes if this should be necessary. If the tubes are welded into the tubesheets and the process fluid conditions are non-fouling, these plugs are unnecessary.

An alternate method of providing access to all tubes for repair and cleaning is to use removable bonnet headers. These designs require gaskets to keep the process fluid from leaking to the atmosphere but may be advantageous for high tube side fouling conditions.

Special header designs may be provided for high tube side pressure conditions. These may be circular headers with individual tubes welded in place or billet type headers with flow passages machined into thick steel sections.

The tube bundle is fabricated as a rigid structure to be handled as an individual assembly. Structural steel side members and tube supports are used for this purpose. Such supports are used beneath the bottom of the tubes to prevent the bundle from sagging; between tube rows to maintain tube spacing and prevent meshing or deformation of the fins; and across the top row of tubes to keep the tubes in proper position. The supports are spaced evenly along the bundle length at intervals not exceeding about 1.5 meters.

Fan and Drive Assemblies

Fans are used which correspond to the dimensions of the tube bundle and the performance requirements for the heat exchanger. Normally, the fan diameter is approximately equal to the bundle width, although smaller diameters may be used. For square, or nearly square bundles, one fan is used. For long rectangular bundles, a number of fans operating in parallel may be used.

The operating point of a fan and tube bundle combination, is determined by matching the fan operating characteristics (see Figure 3.5a) with the air side pressure loss for air flowing through the bundle (Fig. 3.5b). Typically, the process fluid operating conditions are specified and only the air side inlet temperature (that is, design ambient temperature) is given. The air side flowrate is determined by superimposing the two independent curves as shown in Figure 3.5c. The intersection of the two curves indicates the actual air flow rate which can be expected per fan.

Fans are of axial flow design which move relatively large volumes of air at low pressure. In order to minimize air recirculation and improve fan efficiency, fan blades are set within orifice rings which provide close radial clearance between the ring and the blade tips. The ring often has a contoured shape to provide a smooth entrance condition for the air. This minimizes air turbulence at this point which also helps to reduce noise generated by the fan.

Rotating at high speeds, the fan blades must be balanced to insure that centrifugal forces are not transmitted through the fan shaft to the drive or to the supporting structure. An unbalanced blade could result in severe vibration conditions.

Blades are frequently made of aluminum, but other metals and plastics have been used. Consideration of maximum operating temperature must be given when using the plastic blades. Where corrosion is possible, blades can be coated with

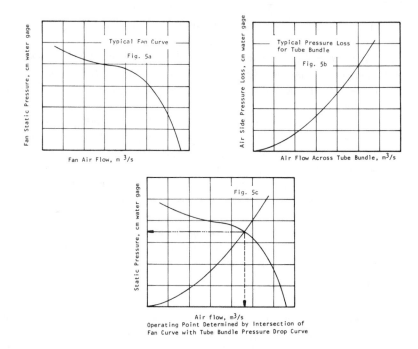

Figure 3.5 Determination of fan operating point.

epoxies or other suitable protective material.

Smaller diameter fans, up to about 1.5 or 2 meters in diameter can be direct driven with electric motors. Larger diameter fans are usually indirectly driven by electric motors or steam turbines using V-belts or gears. V-belt drives are often limited to fan diameters of about 3 meters and less and motors not exceeding 30 hp. For larger motors and larger diameter fans, right angle gear drives are used.

Indirectly driven fans can offer the advantage of speed variation such that as the air cooler heat load varies, the volume of cooling air can also be varied. The "fan laws" which relate speed to fan performance show that reducing speed can also reduce power consumption:

Air flow, m^3/sec	α rpm
Static pressure, $cm \cdot H_2O$	α rpm^2
Horsepower required	α rpm^3

Another benefit in reducing fan speed is reduction of noise generated by the fan. Noise control is becoming increasingly important in industry (13). The noise reduction in decibels may be approximated by: (14)

$$\Delta db = 50 \log (rpm_1 / rpm_2) \tag{6}$$

An alternate way of controlling fan performance is through the use of adjustable pitch fan blades. These may be manually adjustable or automatically adjusted through the use of pneumatic controls. The pitch angle of the blades directly affects the air flow capacity and power requirements of the fan.

The fan may be designed for either forced air flow or induced air flow (see Figure 3.6). In forced-flow installations, the fan blows ambient air across the tube bundle. Induced-draft fans draw the air across the bundle. Therefore, the fan blades are in contact with the heated air coming off the heat exchanger. This situation gives a power advantage for the forced draft design.

Figure 3.6a Large, forced-draft, air-cooled heat exchanger. 21' wide, 44' long with three 13' diameter fans. Including three heat exchanger sections for cooling glycol-water. (Courtesy of Ecodyne, MRM Division.)

Figure 3.6b Induced-draft, air-cooled heat exchanger. Cooling air is pulled through heat exchanger below by fans above heat exchanger. Motors and fan drives are below heat exchanger for easier servicing as well as improved motor, drive belt and bearing life. The cooler above was 12' wide by 30' long with two 10' diameter fans driven by 25 HP motors. (Courtesy of Ecodyne, MRM Division.)

Power requirements for the fans may be estimated from the volume of air required and the total pressure which the fan must deliver in order to move that air across the tube bundle:

$$HP = \frac{\text{Air flow } (m^3/s) \times TP_{fan} \text{ (cm } H_2O)}{7.62 \times \text{Fan Efficiency} \times \text{Drive Efficiency}} \qquad (7)$$

The total pressure of the fan is the sum of the static pressure loss of the air flowing across the tube bundle plus the velocity pressure of the air moving through the fan. Static pressure losses are of the order of 0.5 cm to 3 cm water gage while fans are usually designed for velocity pressure of about 0.25 cm water gage.

The actual volumetric flowrate of air, for a given mass flowrate, is directly proportional to the absolute temperature of the air. Thus, the approximate ratio of power requirements for induced-draft and forced-draft fans is given by:

$$\frac{HP \text{ induced}}{HP \text{ forced}} = \frac{Ta_o}{Ta_i} \qquad (8)$$

Fan efficiencies in Equation (7) are about 65% while drive efficiencies are 95% or better.

This power advantage for forced-draft designs may prove to result in a more economical heat exchanger. Since the fan is close to the ground, structural costs may be less with the drive assembly located at ground level.

However, induced-draft air cooled heat exchangers offer the advantage of better air distribution across the bundle due to relatively low air velocities approaching the tubes. Furthermore, the air exit velocities of induced-draft heat exchangers are much higher than a forced-draft design. Thus, the possibility of recirculating hot discharge air is less for the induced-draft. When cooling the process fluid to a temperature close to the inlet ambient air temperature, this may be of particular importance.

Air Plenum Chamber

The velocity of the air flowing through the fan can be as much as 3 to 4 times the velocity across the face of the tube bundle. Also, the air coming from the circular shape of the fan must be distributed across the square or rectangular shape of the bundle. The air plenum chamber is intended to make this velocity and shape transition such that the distribution of air is uniform across the bundle. Common practice is to install the fan in a chamber such that the distance from the first row of the tube bundle to the fan is about one-half the fan diameter.

The plenum chamber design may be a simple box shape formed by flat sides and bottom, or curved transition sections may be used to obtain a tapered smooth transition from the rectangular bundle to the circular fan. Either design may be used for forced-draft or induced-draft air cooled heat exchangers.

Structural Assembly

The structural assembly of the air cooled heat exchanger is strongly dependent upon the particular plant site requirements of the user. Taken into account should be mechanical loads upon the heat exchanger structure due to its own weight, of course, but other loadings such as wind loads, impact loads, nozzles loading and seismic forces must be considered.

The presence of equipment beneath the air cooled heat exchangers may require particular designs. Safety considerations may call for fencing or fan guards. Environmental factors could indicate the need for louvers, hail screens, or other protective devices. In addition, the physical location of the heat exchangers may require ladders, platforms, railings, safety cages and other miscellaneous items which the user will require.

PERFORMANCE OF AIR COOLED HEAT EXCHANGERS

The thermal performance of an air cooled heat exchanger is determined in much the same fashion as other types of heat transfer equipment (15).

The basic equation for heat transfer is

$$Q = U_o A_o \ \text{MTD} \qquad\qquad (9)$$

where the heat transfer, Q, Joules/sec, removed from the process fluid by the cooling air is a function of the overall heat transfer coefficient, U_o, Joules/sec-$m^2 \cdot {}^\circ K$; the total surface area of the tube bundle, A_o, m^2; and the mean temperature difference, MTD, ${}^\circ K$ between the process fluid and the cooling air.

Determination of the surface area, A_o, was discussed earlier for finned tubes and may be calculated by Equations (1) to (4) or other expressions for different tube and fin types.

Calculation of the overall heat transfer coefficient, U_o, is done by adding together the series resistances to heat transfer such that:

$$U_o = \left(\frac{1}{h_o} + r_{mo} + r_{io} + \frac{1}{h_{io}} \right)^{-1} \qquad\qquad (10)$$

The air side effective heat transfer coefficient, h_o, includes the effects of air velocity, tube and bundle geometry and fin efficiency. Evaluation of this term will be discussed in detail later. The term h_{io} is the tube side heat transfer coefficient, h_i, which has been referred to the outside area of the tube by

$$h_{io} = h_i / (A_o / A_i) \qquad\qquad (11)$$

where A_i is the surface area on the inside of the tubes. The tube side convective coefficient may be determined from the literature for flow inside tubes. Typical ranges for tube side heat transfer coefficients are listed below in Table 3.2.

Table 3.2. Typical Tube Side Heat Transfer Coefficients.

	Tube side heat transfer Coefficient, h_i Joules/sec-m²-°K (Btu/hr-ft²-°F)		
CONDENSING:			
Light hydrocarbons	850 - 1700	(150-300)	
Ammonia	1700 - 3400	(300-600)	
Steam	6800 - 8500	(1200-1500)	
GAS COOLING:			
Air or flue gas	55 - 225	(10-40)	
Hydrocarbon gases, low pressure	170 - 280	(30-50)	
Hydrocarbon gases, high pressure	700 - 1100	(125-200)	
LIQUID COOLING:			
Process water	2800 - 5600	(500-1000)	
Light hydrocarbons	850 - 1700	(150-300)	
Fuel Oil	100 - 225	(20-40)	

Metal resistance r_{mo} is calculated for the particular tube metal, wall thickness, and operating temperatures and referred to the total outside surface by:

$$r_{mo} = \frac{(d_r/100) \ln (d_r/d_i)}{2 K_m} \times \frac{A_o}{A_r} \quad (12)$$

where A_r is the bare tube area based upon the root diameter, d_r.

Process side fouling resistance is referred to the outside surface by the area ratio similar to the tube side heat transfer coefficient:

$$r_{io} = r_i \times \left(\frac{A_o}{A_i}\right) \quad (13)$$

where r_i is determined from available standards (12) or by the user as a result of experience with the particular process conditions. Typical fouling resistances would range from 8.8×10^{-5} sec-m²-°K/joule (0.0005 hr-ft²-°F/Btu) for clean service to 7.0×10^{-4} sec-m²-°K/joule (0.004 hr-ft²-°F/Btu) and higher for contaminated or dirty process fluids.

AIR SIDE EFFECTIVE HEAT TRANSFER COEFFICIENT

In evaluating the air side thermal performance of the finned tubes, three areas should be considered.

a. Air side film coefficient, h_a
b. Fin efficiency, η_{eff}
c. Fin to tube bond resistance, r_{bond}

These parameters are combined to determine the air side effective heat transfer coefficient, h_o, by:

$$h_o = \left(\frac{1}{h_a \, \eta_{eff}} + r_{bond} \times \frac{A_o}{A_r} \right)^{-1} \qquad (14)$$

Each parameter will be treated separately.

AIR SIDE FILM COEFFICIENT

Several investigators have studied the evaluation of air side film coefficients (16, 17, 18, 19). The majority of data available is wind tunnel data. However, recent tests have been conducted on industrial size air cooled heat exchangers (20).

The magnitude of the air side film coefficient is a function of the air velocity, the tube diameter, the fin shape, fin height, fin thickness, fin spacing, and the spacing or pitch of the tubes in the tube bundle, as well as the type of tube pitch. As a result, the analyses of the film coefficient is empirically based.

A correlation was developed based upon tests in several banks of finned tubes by Briggs and Young for prediction of the air side film coefficient:

$$h_a = 0.134 \, \frac{k_a}{(d_r/100)} \, R_{ea}^{0.681} \, P_{ra}^{1/3} \left(\frac{l_f}{S} \right)^{-0.2} \left(\frac{t_f}{S} \right)^{-0.1134} \qquad (15)$$

The Reynolds number R_{ea}, is defined using the root diameter of the tube, and the maximum mass velocity of the air, G_a, flowing through the tube bundle:

$$R_{ea} = \frac{G_a \, (d_{r/100})}{\mu} \qquad (16)$$

The mass velocity, G_a, can be obtained from the face velocity of the air as it approaches the tube bundle such that

$$G_a = \frac{\rho \times F_v}{\sigma} \qquad (17)$$

where σ is the ratio of the minimum cross flow area through the bundle to the total tube face area. For typical tube bundles, σ, calculated by

$$\sigma = 1.0 - (d_r + N_f \, t_f \, 1_f)/P_t \qquad (18)$$

is approximately 0.45 to 0.55 such that the maximum velocity through the bundle is approximately twice the face velocity.

Equations (15) through (18) have been evaluated for typical geometries and results are shown in Figure 3.7. Values of h_a, the air side film coefficient are plotted versus face velocity, FV.

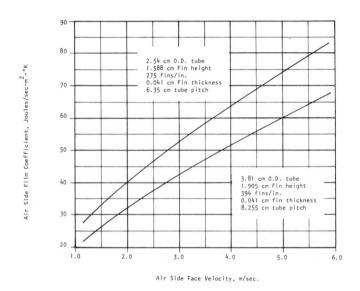

Figure 3.7 Air side film coefficient for typical fin tubes.

Where possible, heat transfer test data for a particular geometry should be used in lieu of generalized heat transfer correlations.

FIN EFFICIENCY

Heat which is transferred from the fin surface to the cooling air must first be conducted through the fin metal. Opposing this conduction of heat is the fin metal resistance which causes a temperature variation along the length of the fin. The effect of this temperature variation, although it actually applies to the temperature difference, is usually used to correct the air side heat transfer coefficient, such that the mean temperature difference of Equation (9) need not be modified.

The derivation of equations used to calculate fin efficiencies are found in the literature for straight fins. For a round fin, Schmidt (21) extends this calculation by developing an expression for an equivalent length, l_e, such that the fin efficiency is given by:

$$\eta_{fin} = \frac{\tanh (ml_e)}{ml_e} \tag{19}$$

where

$$m = \sqrt{\frac{2 h_a}{k_f t_f \times 100}} \qquad (20)$$

and

$$l_e = \left(1 + \frac{t_f}{2 l_f}\right)\left(1 + 0.35 \ln\left[\frac{d_f}{d_r}\right]\right) \qquad (21)$$

Figure 3.8 shows the fin efficiency plotted in graphical form as a function of the parameter $m\, l_e$. The fin efficiency is highly dependent upon the fin material used and the air side film coefficient and can become quite significant in the overall design. Figure 3.9 shows the relative effect of fin materials for typical air cooler tube geometry versus the air side film coefficient. For coated fins, such as soldered or galvanized, the product $K_f t_f$ in Equation (20) would be evaluated by:

$$k_f t_f = k_{\text{fin matl}}\, t_{\text{fin matl}} + k_{\text{coating}}\, t_{\text{coating}} \qquad (22)$$

The fin efficiency, η_{fin}, applies only to the fin surface, itself. The area at the root diameter of the tube can be considered fully effective for heat transfer. Therefore, the effective fin efficiency, η_{eff} is a weighted value:

$$\eta_{\text{eff}} = \frac{A^1_{\text{fin}}\, \eta_{\text{fin}} + (A^1_o - A_{\text{fin}}) \times 1.0}{A^1_o} \qquad (23)$$

The effective fin efficiency is multiplied times the air side film coefficient, h_a, in Equation (14).

BOND RESISTANCE

After the heat from the process fluid travels through the wall of the tube, it must be conducted from the outside surface of the tube to the base of the fin. Resistance to flow of heat at this point is called the bond resistance (22, 23, 24). For bi-metal tubes, the bond resistance would occur between the outer wall of the liner tube, and the inner wall of the outer tube.

When the finned tube is made from two distinct components, the contact area between the two pieces constitute a resistance to heat transfer. This bond resistance is a factor which is dependent upon the method of finned tube manufacture and also the quality of the fabrication. It is a measure of the degree of contact between the tube and fin material.

However, the bond resistance may be influenced by operating temperatures, particularly on the mechanically bonded tubes which rely upon residual stresses to

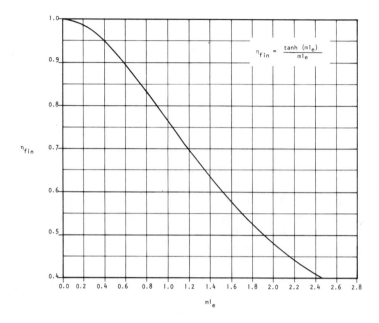

Figure 3.8 Graphical form of fin efficiency.

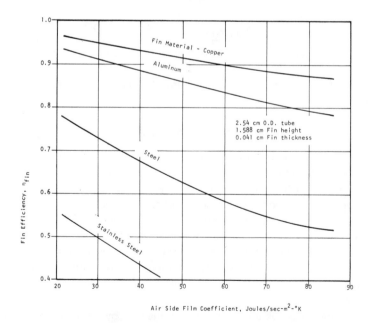

Figure 3.9 Effect of fin materials on fin efficiency.

maintain contact between the fin and tube. As temperatures become high, these stresses may be relieved. Due to unequal thermal expansion of the fin and the tube, an air gap could form between the fin and the tube. This would result in a bond resistance which would degrade the performance of the tube. This is a major reason why these tubes are limited in use to maximum temperatures as indicated earlier.

It could be expected that metallurgically bonded tube would have negligible bond resistance and would be resistant to fin and tube separation as long as operating temperatures were maintained below maximum levels.

Although theoretical analyses have concluded that bond resistances as described above will exist, empirical data for air cooler tubes is limited. However, since heat transfer data is obtained on production tubes, the bond resistance is included in the test results. Therefore, as long as maximum temperature limits are respected, zero bond resistance is assumed.

AIR SIDE PRESSURE DROP

Air side pressure drop in air cooled heat exchangers is important due to the fact that prediction of the quantity of cooling air available is directly dependent upon the prediction of the pressure drop.

Pressure drop through air cooled heat exchangers can be calculated by:

$$\Delta p_a = \frac{G_a^2 / \rho_a}{2g} \left[4 f_{is} \phi_p N_r + 2 \left(\frac{1}{\rho_i} - \frac{1}{\rho_o} \right) \rho_a \right] \tag{24}$$

where the first terms within the bracket are friction losses and the second group represents acceleration losses of the air as it is heated while passing through the heat exchanger.

Robinson and Briggs (25) developed a correlation for the isothermal friction factor, f_{is}, based upon wind tunnel data for various finned tube banks:

$$f_{is} = 9.465 \, Re_a^{-0.316} \left(\frac{P_t}{d_r} \right)^{-0.927} \left(\frac{P_t}{P_d} \right)^{-0.515} \tag{25}$$

where the Reynolds number, Re_a, is based upon the tube root diameter and the maximum velocity through the bundle, (see Equations (16), (17), and (18)).

A correction for variable fluid properties, ϕ_p, is added:

$$\phi_p = \left(\frac{Ta_w}{Ta} \right)^{0.25} \tag{26}$$

As mentioned before for heat transfer results, it is preferable to use test data for a specific tube geometry if possible to evaluate the air pressure drop through the tube bundle of an air cooled heat exchanger.

The above correlations indicate general heat transfer and pressure drop results which could be expected for finned tube bundles with straight circular fins on a staggered pitch. Results for inline tube pitch bundles or tubes on a wide pitch may vary significantly. Similarly, augmented fin designs which are segmented (26), wavy or slotted; rectangular or continuous fin surfaces which join two or more tubes together; and oval tubes with ellipsoid fins are available which would require special correlations. The literature contains correlations for some types. Data for other designs may be obtained from the manufacturer of the air cooled heat exchanger.

MEAN TEMPERATURE DIFFERENCE

In a heat transfer application, the temperature difference is the driving force for the flow of heat. Evaluation of the heat transfer capacity of the air cooled heat exchanger requires calculation of the mean temperature difference between the air side temperatures and the process side fluid temperatures.

The basis for this calculation is the log mean temperature difference (LMTD) for counter flow design. The LMTD is calculated using the terminal temperatures of the two fluids:

$$\text{LMTD} = \frac{(t_i - Ta_o) - (t_o - Ta_i)}{\ln\left(\dfrac{t_i - Ta_o}{t_o - Ta_i}\right)} \tag{27}$$

The outlet air temperature, Ta_o, must be calculated from the heat transfer from the exchanger and from the mass flow of air, such that

$$Ta_o = Ta_i + \frac{Q}{W_a\, Cp_a} \tag{28}$$

For a large number of tube passes, the flow will be essentially counterflow. Thus, the mean temperature difference correction, F_T, is equal to 1. However, if the number of tube side passes is limited by tube side velocity or tube side pressure drop, the value of F_T, will be less than unity. The value of F_T is available in the literature (27) for crossflow heat exchangers, both fluids unmixed such that the mean temperature difference is given by

$$\text{MTD} = \text{LMTD} \times F_T \tag{29}$$

Values of F_T depend upon the relative temperature changes of the air and process fluids and upon the closeness of the approach between the outlet process fluid temperature and the outlet air temperature. If both outlet temperatures are equal,

F_T will be about 0.91 for a single pass design and 0.96 for two passes on the tube side. When temperature cross occurs such that the process fluid outlet temperature is less than the air outlet temperature, low values of F_T can reduce the mean temperature difference significantly.

The LMTD calculation assumes that no phase change occurs as the process side temperature decreases. For condensing service, a heat release curve versus process fluid temperature should be used to accurately determine the effective mean temperature difference.

After determination of the overall heat transfer coefficient, U_o; the total surface area, A_o; and the mean temperature difference, MTD, the overall performance capacity of the heat exchanger is calculated from Equation (9). A comparison of this heat load with required performance determines the adequacy of the air cooled heat exchanger design.

DESIGN CONSIDERATIONS

Several parameters have significant effect upon the design of air cooled heat exchangers and should be considered carefully.

Design Ambient Conditions

a. Specifications of the design ambient air temperature determines the magnitude of the temperature approach of the process fluid to the inlet temperature. This is normally the limiting parameter for heat transfer design of air cooled heat exchangers. An unreasonably high temperature could result in an unnecessarily large surface area requirement. Conversely, if the temperature specified is too low, plant capacity could be seriously affected during warm weather.

The design ambient could be the average daily maximum temperature for the hottest month of the year; or a temperature which would not be exceeded more than 5% of the time during the hottest months of the year.

b. Since density of air is affected by altitude, the altitude of the air cooled heat exchanger site should be specified.

c. Not only should maximum ambient temperatures be specified, but minimum environmental temperatures should be considered. Operation at subfreezing temperatures could result in tube side freezing or formation of ice on the heat transfer surface. In some cases, design for intentional recirculation of warm exhaust air might be warranted (28).

d. The nature of corrosive elements in the air from other equipment operating nearby should be considered when determining materials of construction.

e. Air side fouling is usually not a problem. However, if ambient conditions are known to be poor, surface cleaning of the fins should be anticipated. Steam cleaning, use of pressurized water or air or hand brushing will help to keep surfaces clean and effective.

Process Side Conditions

a. Process fluid operating temperatures and pressures are required for thermal and mechanical design. However, maximum temperatures and pressures to which the heat exchanger would be exposed should also be considered, even if these would only be transient conditions.

b. Fouling resistance should be reasonable values based upon experience. Unnecessarily high values of fouling resistances could result in large uneconomical designs. If fouling is expected to be severe, removable bonnet or removable cover plate headers should be considered.

c. The pressure drop taken through the tube side of the air cooled heat exchanger can greatly influence the overall design of the exchanger since it may affect the number of passes, the tube length and the tube diameter.

These parameters are important in the thermal design of the heat exchanger. Unreasonably low pressure drop specifications can result in an inefficient design.

General Site Characteristics

General requirements due to space limitations, safety precautions, availability of operating and maintenance personnel, and environmental conditions must all be considered in the design of air cooled heat exchangers.

NOMENCLATURE

A_{fin}^1, A_r^1, A_o^1 Surface densities, m^2/m^2 Face Area/tube row of finned surface, bare tube surface and total surface, respectively.

A_o, A_i, A_r Total outside surface, total inside surface, total bare tube surface, respectively, m^2

Cp_a Specific heat of air, Joules/kg-°K

d_f, d_r, d_i Fin diameter, root diameter, and tube inside diameter, respectively, cm

f_{is} Isothermal air side friction factor

F.A. Tube bundle face area, m^2

FT Correction to log mean temperature difference

FV Air face velocity, m/sec

g Gravitational constant, m/sec^2

G_a Air side mass velocity, $kg/sec-m^2$

h_a Air side film coefficient, $Joules/sec-m^2-°K$

h_i Tube side heat transfer coefficient, $Joule/sec-m^2-°K$

h_{io} Tube side coefficient referred to outside surface, $Joule/sec-m^2-°K$

h_o Effective air side heat transfer coefficient, $Joule/sec-m^2-°K$

HP Fan power requirement, horsepower

k_a, k_f, k_m Thermal conductivity of air, fin material and tube material respectively, $Joule/sec-m-°K$

l_e, l_f Effective fin length, and fin height, respectively, cm

L Tube bundle length, m

LMTD Log mean temperature difference, °K

m Heat transfer factor for fin efficiency, cm^{-1}

MTD	Mean temperature difference, $^\circ$K
N_f	Fin spacing, fins/m
N_r	Number of tube rows in bundle in direction of air flow
P_d	Diagonal tube pitch, cm
Pr_a	Air Prandtl number
P_t	Transverse tube pitch, cm
Q	Heat transfer, Joules/sec
r_{bond}	Fin bond resistance, sec-m^2-$^\circ$K/Joule
Re_a	Air side Reynolds number
r_i	Tube side fouling resistance, sec-m^2-$^\circ$K/Joule
r_{io}	Tube side fouling referred to outside surface, sec-m^2-$^\circ$K/Joule
r_{mo}	Tube metal resistance, sec-m^2-$^\circ$K/Joule
rpm	Fan rotational speed, rev/min
S	Space between fins (S $= 100/N_f - t_f$), cm
T_a, Ta_w	Average air temperature, air temperature at tube wall, respectively, $^\circ$K
Ta_i, Ta_o	Inlet and outlet air temperatures, $^\circ$K
t_e, t_f, t_r	Fin thickness: Fin tip, average, and fin base, respectively, cm
t_i, t_o	Process fluid inlet and outlet temperatures, $^\circ$K
TP_{fan}	Total pressure of fan, cm water gage
U_o	Overall heat transfer coefficient, Joule/sec-m^2-$^\circ$K
W_a	Mass flow of air, k$_g$/sec
α	Thermal expansion coefficient, m/m/$^\circ$K
Δdb	Change in noise level, db
ΔL	Change in tube length due to thermal expansion, m
Δt_{amb}	Temperature difference between operating temperature and 294°K
η_{eff}	Effective fin efficiency
η_{fin}	Fin efficiency
μ	Air viscosity, Newton-sec/m^2
ϕ_p	Property variation correction to air side pressure drop
ρ_a, ρ_i, ρ_o	Air densities @ average, inlet and outlet air temperatures, k$_g$/m^3
σ	Ratio of minimum air crossflow area to total face area

REFERENCES

1. Gazzi, L. and Pasero, R., "Process Cooling Systems — Selection," Hydrocarbon Processing, Oct. 1970, pp. 83–90.
2. Kassat, H., "Air Cooled Heat Exchangers in Process Engineering," translation of article appearing in *Chemiker Zeitung*, Heidelburg, 1971, pp. 596–606.
3. Kassat, H., "Examples of Use and Design of Air-Cooled Closed Cooling Water Systems," translation of article appearing in *Technische Mitteilungen*, 62nd yr, 1969, pp. 441–448.
4. Rubin, F. L., "Design of Air Cooled Heat Exchangers," Chemical Engineering, Oct. 31, 1960, pp. 91–96.
5. Schoonman, W., "Air Cooled Optimization by Computer," ASME Symposium on Air-Cooled Heat Exchangers, 1964, pp. 86–102.

6. Schoonman, W., "Air Cooled Steam Condensers for Power Plants," Society of Mechanical Engineers Symposium, Decentralization of Energy Production in Relation to Available Water Resources, Sydney, Australia, Sept. 1970.

7. Williams, C. L. and Damron, R. D., "Chemical Plant Air Cooled Exchanger Economics and Design," ASME Symposium on Air-Cooled Heat Exchangers, 1964, pp. 103–125.

8. Brown, J. W. and Berkly, G. J., "Heat Exchangers in Cold Service — A Contractor's View," Chemical Engineering Progress, Vol. 70, No. 7, July 1974, pp. 59–62.

9. Weierman, C., "Finned Tubes Can Lower Heat-Transfer Costs," The Oil and Gas Journal, Nov. 3, 1975, pp. 64–72.

10. Gardner, K. A., "Rational Design Temperature Limits for Interference — Fit Finned Tubing," ASME Symposium on Air-Cooled Heat Exchangers, 1964, pp. 1–20.

11. "Air-Cooled Heat Exchangers for General Refinery Services," American Petroleum Institute, API Standard 661, First Edition, August 1968.

12. "Standards of Tubular Exchanger Manufacturers Association," Tubular Exchanger Manufacturers Association, Inc., Fifth edition, 1968.

13. Cheremisinoff, P. N. and Cheremisinoff, P. P., Editors, *Industrial Noise Control Handbook*, Ann Arbor Science, 1977.

14. Engstrom, S. P., "Silencing Noisy Machines," Machine Design, Jan. 26, 1978, pp. 84–89.

15. Kreith, F., *Principles of Heat Transfer*, Second edition, International Textbook Company, 1967.

16. Briggs, D. E. and Young, E. H., "Convection Heat Transfer and Pressure Drop of Air Flowing across Triangular Pitch Banks of Finned Tubes," CEP Symp., Ser. No. 41, *59*, 1963, p. 1.

17. Cox, B., "Heat Transfer and Pumping Power Performance in Tube Banks—Finned and Bare," ASME-AICHE Heat Transfer Conference, 1973, Paper N 73 — HT-27, 10 p.

18. Rabas, T. J. and Eckels, P. W., "Heat Transfer and Pressure Drop Performance of Segmented Extended Surface Bundles," AICHE-ASME, 15th National Heat Transfer Conference, 1975.

19. Weierman, C., "Correlations Ease the Selection of Finned Tubes," The Oil and Gas Journal, Sept. 6, 1976, pp. 95–100.

20. Cowan, G. C., Dell, F. R., and Stinchombe, R. A., "Aerodynamics and Heat Transfer Performance of an Industrial Air-Cooled Heat Exchanger," 5th Int. Heat Transfer Conference, V5, 1974, pp. 155–159.

21. Schmidt, T. E., "Heat Transfer Calculations for Extended Surfaces," J. of the ASRE, 1949, pp. 351–357.

22. Gardner, K. A. and Carnovos, T. C., "Thermal Contact Resistance in Finned Tubing," Journal of Heat Transfer, Nov. 1960, pp. 279–293.

23. Kraus, A. D., "The Effect of Bond Resistance on the Overall Efficiency of a Heat Exchanger Core," 5th International Heat Transfer Conf., V5, 1974, pp. 165–169.

24. Young, E. H. and Briggs, D. E., "Bond Resistance of BiMetallic Finned Tubes," Chemical Engineering Progress, Vol. 61, No. 7, July 1965, pp. 71–79.

25. Robinson, K. K. and Briggs, D. E., "Pressure Drop of Air Flowing Across Triangular Pitch Banks of Finned Tubes," AICHE, 8th National Heat Transfer Conference, 1965.

26. Weierman, C., "Pressure Drop Tests on Welded Plain and Segmented Finned Tubes in Staggered and Inline Layouts," AICHE-ASME, 16th National Heat Transfer Conference, 1976.

27. Bowen, R. A., Mueller, A. C., and Nagle, W. M., "Mean Temperature Difference in Design," Transactions of the ASME, Vol. 62, 1940, pp. 283–294.

28. Shipes, K. V., "Air-Cooled Exchangers in Cold Climates," Chemical Engineering Progress, Vol. 70, No. 7, July 1974, pp. 53–58.

CHAPTER 4

SHELL AND TUBE HEAT EXCHANGERS

J. H. KISSEL
American Standard Inc.
Heat Transfer Division
Buffalo, NY

INTRODUCTION

This chapter will consider shell and tube heat exchangers, more specifically, conventional commercial shell and tube heat exchangers used for water, oil or gas heating or cooling operation.

It will cover applicable Codes and Standards and describe the typical types and configurations of heat exchangers with respective applications. There will be a brief treatise on thermal rating, mechanical design, problem areas and proper maintenance of the heat exchanger to insure long service life.

CODES AND STANDARDS

There are myriads of Codes and Standards that dictate heat exchanger design. Federal, state and insurance company codes play a part and are chiefly concerned with product safety which of course is also the primary concern of the heat exchanger designer. In conjunction with this, customer specifications often govern external design considerations, testing, mechanical and thermal requirements.

Codes or sets of rules, however, offer only general guidance and set minimum standards. The success or failure of a given design still is the ultimate responsibility of the designer and only proper application of his good engineering judgment results in a long life, cost effective heat exchanger.

For commercial installation, the Boiler and Pressure Vessel Code of the American Society of Mechanical Engineers (ASME) (1) generally apply. These rules establish the design, fabrication and inspection during construction of boilers and pressure vessels. A shell and tube heat exchanger is considered a pressure vessel and is therefore covered by these rules. The ASME Code presently consists of eleven sections. Section III Division I covers Nuclear Components and Section VIII covers Pressure Vessels. The objective of the rules is to afford reasonably certain protection of life and property and to provide a margin for deterioration in service so that a reasonably long safe period of usefulness can be attained.

A number of shell and tube heat exchanger manufacturers have formed the Tubular Exchanger Manufacturers Association (TEMA) (2) and have issued standards for the mechanical design, fabrication, installation, operation, and maintenance of heat exchangers. Included in these standards are thermal, physical property, and other useful data for the designer. The emphasis in these standards

59

concern the mechanical features of heat exchangers and the standards now include a recommended good practices section which gives the designer additional information and guidance relative to the design of heat exchangers.

The TEMA standards have three basic classifications; R, C and B.

TEMA Class R

The TEMA Mechanical Standards for Class-R heat exchangers are used for generally severe requirements of petroleum and related processing applications. Equipment fabricated in accordance with these standards is designed for safety and durability under the rigorous service and maintenance conditions of such applications.

TEMA Class C

The TEMA Mechanical Standards for Class-C heat exchangers are used for commercial and general-purpose applications. Equipment designed in accordance with these standards is designed for maximum economy and overall compactness consistent with the safety and service requirements of such applications.

TEMA Class B

These standards are used for chemical process service and again are designed for maximum economy and overall compactness consistent with the Safety and Service requirements of such applications.

HEAT EXCHANGER TYPES AND CONFIGURATIONS

The shell and tube heat exchanger is basically what the name implies. It consists of a shell, usually a circular cylinder, with a large number of tubes attached to an end plate and arranged in a fashion where two fluids can exchange heat without the fluids coming in contact with one another.

The most common types of heat exchangers and the nomenclature of heat exchanger components is shown in the TEMA Standards.

Figure 4.1 depicts the outline of the more common types of shell and tube heat exchangers.

Each of these types serves a specific need and proper application of the type is a necessary criteria for operational longevity.

Table 4.1 is a useful guide for selection of the proper heat exchanger.

THERMAL RATING

The thermal design of shell and tube heat exchangers involves a number of variables and prediction of the heat transfer capability is rather complex.

Process Equipment Series Volume 2

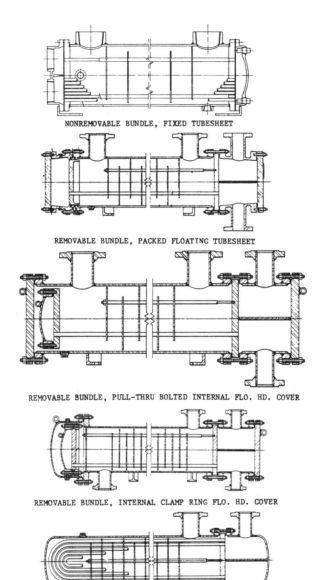

NONREMOVABLE BUNDLE, FIXED TUBESHEET

REMOVABLE BUNDLE, PACKED FLOATING TUBESHEET

REMOVABLE BUNDLE, PULL-THRU BOLTED INTERNAL FLO. HD. COVER

REMOVABLE BUNDLE, INTERNAL CLAMP RING FLO. HD. COVER

REMOVABLE BUNDLE, "U" TUBE

Figure 4.1

Table 1. Heat Exchanger Applications.
(Use this guide and the following charts to determine the heat exchanger you need.)

CONSTRUCTION	ADVANTAGES	LIMITATIONS	SELECTION TIPS
NONREMOVABLE BUNDLE, FIXED TUBESHEET	1. Usually less costly than removable bundle heat exchangers. 2. Provides maximum heat transfer surface per given shell & tube size. 3. Provides multi-tube pass arrangements.	1. Shell side can be cleaned only by chemical means. 2. No provision* to correct for differential thermal expansion between the shell and tubes. *Exception-expansion joint can be installed as reqd.	1. For lube oil & hydraulic oil coolers, put the oil thru the shell side. 2. Corrosive or high fouling fluids should be put thru the tube side. 3. In general, put the coldest fluid thru the tube side.
REMOVABLE BUNDLE, PACKED FLOATING TUBESHEET	1. Floating end allows for differential thermal expansion between the shell and tubes. 2. Shell side can be steam or mechanically cleaned. 3. Bundle can be easily repaired or replaced. 4. Less costly than full internal floating heat type construction. 5. Maximum surface per given shell and tube size for removable bundle designs.	1. Shell side fluids limited to nonvolatile and/or non-toxic fluids, i.e. lube oils, hydraulic oils. 2. Tubes expand as a group, not individually (as in a U-tube unit) therefore sudden thermal shocking should be avoided. 3. Packing limits design pressure and temperature.	1. For lube oil or hydraulic oil coolers put the oil thru the shell side. 2. For air intercoolers and aftercoolers on compressors put air thru the tube side. 3. Put hot shell side fluid thru at stationary end (to keep temperature of packing as low as possible).

(Continued)

Table 1. Heat Exchanger Applications (Continued).

CONSTRUCTION	ADVANTAGES	LIMITATIONS	SELECTION TIPS
REMOVABLE BUNDLE, PULL-THROUGH BOLTED INTERNAL FLOATING HEAD COVER	1. Allows for differential thermal expansion between the shell and the tubes. 2. Bundle can be removed from shell for cleaning or repairing without removing the floating head cover. 3. Provides multi-tube pass arrangements. 4. Provides large bundle entrance area. 5. Excellent for handling flammable and/or toxic fluids.	1. For a given set of conditions, it is the most costly of all the basic types of heat exchanger design. 2. Less surface per given shell and tube size than removable bundle internal clamp ring type floating head cover.	1. If possible, put the fluid with the lowest heat transfer coefficient thru the shell side. 2. If possible, put the fluid with the highest working pressure thru the tube side. 3. If possible, put the high fouling fluid thru the tube side.
REMOVABLE BUNDLE, INTERNAL CLAMP RING TYPE FLOATING HEAD COVER	1. Allows for differential thermal expansion between the shell and the tubes. 2. Excellent for handling flammable and/or toxic fluids. 3. Provides multi-pass arrangements.	1. Shell cover, clamp-ring and floating head cover must be removed prior to removing the bundle.. 2. More costly than fixed tube sheet or U-tube heat exchanger design.	1. If possible, put the fluid with the lowest heat transfer coefficient thru the shell side. 2. If possible, put the fluid with the highest working pressure thru the tube side. 3. If possible, put the high fouling fluid thru the tube side.

(Continued)

Table 1. Heat Exchanger Applications (Continued).

CONSTRUCTION	ADVANTAGES	LIMITATIONS	SELECTION TIPS
REMOVABLE BUNDLE, U-TUBE	1. Less costly than floating head or packed floating tubesheet designs. 2. Provides multi-tube pass arrangements. 3. Allows for differential thermal expansion between the shell and the tubes, as well as between individual tubes. 4. High surface per given shell and tube size. 5. Capable of withstanding thermal shock.	1. Tube side can be cleaned only by chemical means. 2. Individual tube replacement is difficult, thus high maintenance costs. 3. Cannot be made single pass on tube side, therefore, true counter-current flow is not possible. 4. Tube wall at U-bend is thinner than at straight portion of tube. 5. Draining tube side difficult in vertical, (head-up) position.	1. For oil heaters, wherever possible put steam thru the tube side, to obtain the most economical size.

(Concluded)

64

There are many text books that describe the fundamental heat transfer relationships but few discuss the complicated shell side characteristics. On the shell side of a shell and tube heat exchanger, the fluid flows across the outside of the tubes in complex patterns. Baffles are utilized to direct the fluid through the tube bundle and are designed and strategically placed to optimize heat transfer and minimize pressure drop. A measure of the complexity of predicting shell side heat transfer can be obtained by considering the path of shell side fluid flow.

The flow is partially perpendicular and partially paralleled to the tubes. It reverses direction as it travels around the baffle tips and the flow regime is governed by tube spacing, baffle spacing and leakage flow paths. Throughout the fluid path, there are a number of obstacles and configurations which cause high localized velocities. These high velocities occur at the bundle entrance and exit areas, in the baffle windows, through pass lanes and in the vicinity of tie rods which secure the baffles in their proper position. In conjunction with this, the shell side fluid generally will take the path of least resistance and will travel at a greater velocity in the free areas or by-pass lanes than it will through the bundle proper where the tubes are on a closely spaced pitch. All factors considered, it appears a formidable task to accurately predict heat transfer characteristics of a shell and tube exchanger.

The problem is further complicated by the manufacturing tolerances or clearances that are specified to allow assembly and disassembly of the heat exchanger. It is improbable that these clearances will all accumulate to either the positive or negative side so it is customary to compute heat transfer relationships on the basis of average clearances.

The various paths of fluid flow through the shell side of a segmental baffle heat exchanger has been considered by Tinker (3). The various streams illustrated in Figure 4.2 are defined in Tinker's work as follows:

Stream	Description
A.	Leakage stream through the annular spaces between tubes and baffle holes of one baffle.
B.	Cross flow stream through the heat transfer surface between successive baffle windows. It will be noted that this stream is made of B_1 (a portion of fluid passing through baffle windows) plus portions of the A stream.
C.	By-pass stream on one side of tube nest flowing between successive baffle windows.
E.	Leakage stream between shell and edge of one baffle.

The by-pass area C between the bundle and shell can be reduced by using dummy tubes, sealing strips, or tie rods with seal strip baffles. The dummy tubes do not pass through the tubesheets and can be located close to the inside of the shell. The sealing strips extend from baffle to baffle in a longitudinal direction and effectively channel the fluid across the tubes to minimize turbulence and heat

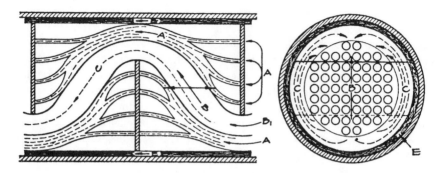

Figure 4.2 Leakage path streams (T. Tinker).

transfer. On some fixed tubesheet designs, the outer tubes are in close proximity to the inside of the shell so that by-pass is minimal and no by-pass elimination is necessary.

There are a number of techniques that can be employed to reduce the flow in areas A and E. Tight tolerances are often employed and some manufacturers use a punched collar baffle where the tube holes in the baffle have a small precision collar which minimizes clearances between tube and tubehole with the added benefit of good tube support (see Figure 4.3). The baffles are sometimes welded at its periphery to the shell to completely eliminate by-pass of the E stream. Each of these techniques is effective but are governed by the trade off of increased efficiency versus added cost.

There are many good methods for predicting commercial shell-and-tube heat exchanger shell side heat transfer characteristics. Among these is Tinker's method and is responsive to the major effects of unit size, tube diameters and pitch, baffle spacing and cut and particularly the principal leakage streams as described above. Tinker's work was based on research data and is practical in nature. The work was accomplished in the pre-computer era and certain resistance factor approximations and assumptions were used to derive simple and useable formulas. Much more sophisticated computer techniques are employed today with some differences noted at the high and low values of Reynolds numbers.

In Tinker's work, the heat transfer group (B_o) for flow across tubes is plotted as a function of Reynolds number for various pitch patterns. To illustrate his work, only the triangular pitch pattern for shell side heat transfer of the baffled portion will be considered (Figure 4.4).

The shell side heat transfer coefficient for the baffled region of the tube bundle is:

$$h_{ob} = \frac{16.1}{d_2} \, B_o \, K \left(\frac{CZ}{K} \right)^{1/3} \phi \tag{1}$$

where:

h_{ob} = shell side heat transfer coefficient (Btu/hr-ft^2 °F)

d_2 = tube outside diameter (in.)

B_o = heat transfer factor Figure 4.4

K = fluid thermal conductivity (Btu/hr-ft^2 °F per ft)

C = fluid specific heat (Btu/lb°F)

Z = fluid viscosity at its mean temperature (centipoises)

Zw = fluid viscosity at tube wall (centipoises)

ϕ = $(Z/Zw)^{0.14}$

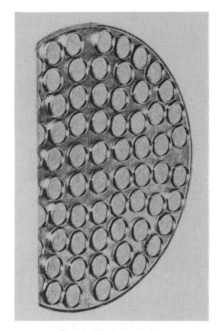

Figure 4.3 Collar baffle.

To use Figure 4.4, the following terms apply:

D_1 = shell inside diameter (in.)

D_2 = baffle outside diameter (in.)

D_3 = bundle diameter, outer tube limit (in.)

l_3 = baffle spacing (in.)

H = height of baffle cut (in.)

P = tube pitch (in.)

d_1 = tube hole diameter in baffle (in.)

N_H = rating constant for heat transfer calculation

For a typical case of a fixed tubesheet unit:

$$\frac{D_1}{D_3} = 1.02 \qquad \frac{d_1 - d_2}{d_2} = .02 \qquad \frac{D_1 - D_2}{D_1} = .0048 \qquad (2)$$

P/d_2 $\overline{D_1/l_3}$	N_H		
	1.25	1.3	1.4
1	.10	.10	.07
2	.18	.16	.13
4	.34	.28	.22

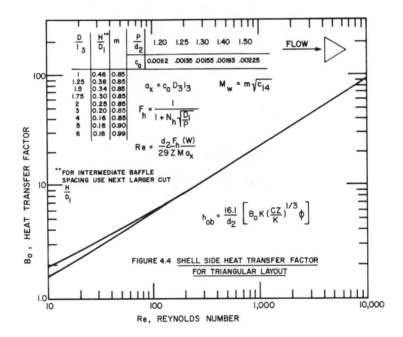

FIGURE 4.4 SHELL SIDE HEAT TRANSFER FACTOR
FOR TRIANGULAR LAYOUT

Figure 4.4

For other clearances and pitch configurations, the N_H value will change in turn changing the shell side coefficient.

There are, of course, many other calculation methods for the prediction of the shell side flow pressure drop and heat transfer. Some of these are very sophisticated such as the Stream Analysis Method (4) which is a widely used proprietary tool of

Heat Transfer Research, Inc. It is based on an analysis of the individual stream components and the interaction of constructional parameters with solutions made possible by use of high speed digital computers.

Since the designer has these procedures at his disposal, he is in a much better position to optimize his design by varying baffle type and cut, pitch and flow orientation entering and leaving the heat exchanger.

The baffle type is an important consideration in heat exchanger design and primarily serves to provide structural support for the tubes and to direct the shell side fluid. The most common type is the segmental baffle and can be single, double or triple segmented depending on the shell side pressure drop, and heat transfer considerations. These baffles are oriented to provide either a side-to-side or up and down motion to the fluid and can be utilized in a split flow or multi pass shell arrangement.

A split flow arrangement is one where the shell side fluid enters the middle of the shell, splits and exits at the ends of the exchanger. The obvious advantage is that only half of the flow crosses any baffled zone in the heat exchanger thereby reducing velocity and pressure drop.

A multi shell pass arrangement utilizes a longitudinal baffle to divide the shell into two or more sections and can serve to increase shell side flowrate thereby increasing velocity and heat transfer.

Figure 4.5 depicts some typical baffle and shell arrangements.

Any combination is possible and for a given set of design considerations, one design may have cost effective advantage over the others. Generally, the more complex the design becomes, the more difficult it is to determine the effect of the shell side leakages and consequently the shell side heat transfer. A common problem in heat exchanger design is to accurately predict the leakage past the longitudinal baffle in a multi pass shell exchanger.

MECHANICAL DESIGN

A shell and tube heat exchanger is a pressure vessel and Section VIII Division 1 of the ASME Code contains numerous rules for the design of the pressure retaining parts. Formulas are presented in the Code for the calculation of flanges, covers, pipe and heads with reference to an allowable stress which is a percentage of the yield or ultimate stress of the material. These formulas cover the conventional geometric shapes such as cylinders and circular plates although rules are in progress for rectangular vessels and items germane to heat exchangers such as tubesheets, expansion joints and tube joints.

A thorough heat exchanger design analysis will include consideration of factors such as deadweight loading, external piping loads, wind loads, vibration, seismic and thermal transients. These loads will affect the nozzles, shells, headers, component supports and internals of the heat exchanger. Very often, these loads are small and can be neglected in the design but on occasion a start-up or shut-down condition

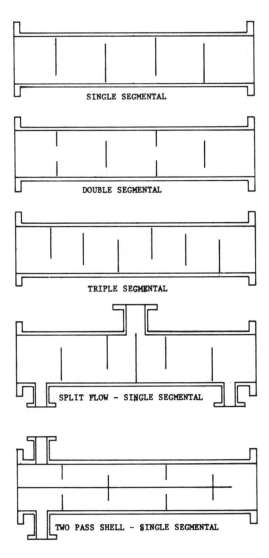

Figure 4.5 Typical baffle and shell arrangement.

exists where thermal stresses or external loads can be high. If such a condition exists, it is the responsibility of the user to alert the designer so that he can consider these loads in the design.

The analytical approach to stress analysis and mechanical design is slowly yielding to powerful numerical techniques such as finite element analysis. Still appropriate, however, are reference sources such as R. J. Roark, *Formulas for Stress and*

Strain (5), and S. Timoshenko and S. Woinowsky-Kreiger, *Theory of Plates and Shells* (6) since they contain basic analytical approaches for almost all of the geometric shapes encountered in heat exchangers. Skillful manipulation of these formulas with the help of super position can generally lead to the solution of a design problem. There are supplementary analytical approaches for specific problems which are widely used and accepted by the industry. Foremost of these are Zick's (7) work on saddle supports and Welding Research Council Bulletin #107 (8) for the analysis of local stresses in a shell due to external loads applied to a nozzle or other types of attachments. The TEMA standards contain a good analysis of fixed tubesheet heat exchangers and is based on the work of Karl Gardner. ASME has published (9, 10) a number of volumes which summarizes the state of the art in pressure vessel design and has a collection of the seminal papers with a selective bibliography for each of the topics discussed.

There are many other excellent reference sources that the designer has at his disposal. The real problem is that many of these sources are highly technical and difficult to interpret and apply.

There seems to be a vast middle ground between the innovative mechanical designer who possesses more intuition than technical background and the highly analytical person who can manipulate higher mathematics to his bidding. The designer would like to see simple useable formula but the mathematician finds the chore to convert the mathematical derivation to simple rules or formulas is often a greater task than performing the derivation in the first place.

The way to bridge this middle ground is with the use of the computer, and the computer has spawned the use of various numerical techniques such as finite element analysis. This is a very powerful tool and odd geometric shapes can be analyzed as readily as the conventional rings, shells and plates. A knowledge of higher mathematics is no longer a necessity and three dimensional analysis is possible. The main disadvantage of this method is the cost for computer time. However, computer costs are rapidly decreasing and the art of using finite elements is improving so that this approach will undoubtedly obsolete the analytical one in the near future.

In summary, the mechanical designer has many analytical and numerical techniques at his disposal for conventional designs. Moreover, these techniques are becoming quite refined and more easily implemented.

PROBLEM AREAS AND MAINTENANCE

Shell and tube heat exchangers are generally designed with a certain degree of conservatism from both the thermal and mechanical design aspects. From a thermal design viewpoint, the conservatism arises from excessive surface to accommodate fouling in service. From a mechanical design viewpoint, design procedures generally employ allowable stresses which are based on a factor of safety. But, even so, shell and tube heat exchangers experience problems in service.

One of these problems concerns fouling of either the tube side or shell side of the heat exchanger. Fouling is an accumulation of scale or dirt on the tube surface thereby adding a resistance to heat transfer. It is very difficult to accurately predict the degree of fouling for a specified service period. There are minimal documented test results on this subject and the results are seldom applicable because of the number of variables in a fouling study. It is indeed a fortunate user who can rely on past performance of the same or similar equipment and specify the proper amount of excess surface required to offset the amount of fouling. For most applications, the degree of fouling is strictly an estimate and the probability is that the heat exchanger is either inadequate or over surfaced.

Once the tubes are fouled, they can be either mechanically or chemically cleaned. Generally the tube side presents no particular problem and straight tubes can be easily wire brushed. U-tubes are difficult to clean mechanically and are

Table 4.2. Recommended Start-Up and Shut-Down Procedures.

Caution: Every effort should be made to avoid subjecting the unit to thermal shock, overpressure, and/or hydraulic hammer, since these conditions may impose stresses that exceed the mechanical strength of the unit or the system in which it is installed which may result in leaks and/or other damage to the unit and/or system.

Heat Exchanger Type of Construction	Fluid Location & Relative Temp.				Start-Up Procedure	Shut-Down Procedure
	Shell side		Tube side			
	Type of Fluid	Rel. Temp.	Type of Fluid	Rel. Temp.		
Fixed Tubesheet (Non-Removable Bundle)	Liquid	Hot	Liquid	Cold	Start both fluids gradually at the same time.	Shut down both fluids gradually at the same time.
	Condensing Gas (i.e.,Steam)	Hot	Liquid or Gas	Cold	Start hot fluid first, then cold fluid.	Shut down cold fluid first, then hot fluid.
	Gas	Hot	Liquid	Cold	Start cold fluid first, then hot fluid.	Shut down cold fluid gradually, then hot fluid.
	Liquid	Cold	Liquid	Hot	Start both flows gradually at the same time.	Shut down both fluids gradually at the same time.
	Liquid	Cold	Gas	Hot	Start cold fluid first, then hot fluid.	Shut down hot fluid first, then cold fluid.
U-Tube Packed Floating Head Packed Floating Tubesheet Internal Floating Head (All these types have Removable Bundles)	Liquid	Hot	Liquid	Cold	Start cold fluid first, then start hot fluid gradually.	Shut down hot fluid first, then cold fluid.
	Condensing Gas (i.e. Steam)	Hot	Liquid or Gas	Cold	Start cold fluid first, then start hot fluid gradually.	Shut down cold fluid first, then shut down hot fluid gradually.
	Gas	Hot	Liquid	Cold	Start cold fluid first, then start hot fluid gradually.	Shut down hot fluid first, then cold fluid.
	Liquid	Cold	Liquid	Hot	Start cold fluid first, then start hot fluid gradually.	Shut down hot fluid first, then cold fluid.
	Liquid	Cold	Gas	Hot	Start cold fluid first, then start hot fluid gradually.	Shut down hot fluid first, then cold fluid.

General Comments:
1) In all start-up and shut-down operations, fluid flows should be regulated so as to avoid thermal shocking the unit regardless of whether the unit is of either a removable or non-removable type construction.
2) For fixed tubesheet (non-removable bundle) type units where the tube side fluid cannot be shut down, it is recommended that (1) A bypass arrangement be incorporated in the system, and (2) The tube side fluid be bypassed before the shell side fluid is shut down.

generally used where fouling is expected to be minimal. The shell side of the heat exchanger is more difficult to clean, particularly for closely spaced staggered types of tube bundles. Many users specify square or rectangular pitch tube arrangement and removable bundle construction where excessive shell side fouling is expected.

Another serious problem in heat exchangers is corrosion. Severe corrosion can and does occur in tubing and very often with common fluids such as water. Proper material selection based on a full analysis of the operating fluids, velocities and temperatures is mandatory. Very often, heavier gauge tubing is specified to offset the effects of corrosion but this is only a partial solution. This should be followed by proper start-up, operating and shut-down procedures. Many heat exchangers use water on the tube side as the cooling medium and compatible copper alloy tubing and still experience corrosion problems. Invariably, this can be traced to some part of the cycle where the water was stagnant or circulated at extremely low velocity.

Most problems with heat exchangers occur during initial installation or shortly thereafter. Improper installation or misalignment can create excessive stresses in supports or nozzles or cause damage to expansion joints or packed joints. On initial start-up and shut-down the heat exchanger can be subjected to damaging thermal shock, over pressure or hydraulic hammer. This can lead to leaky tube-to-tubesheet joints, damaged expansion joints or packing glands because of excessive axial thermal expansion of the tubes or shell. Excessive shell side flowrates during the "shake down" can cause tube vibrations and catastrophic failure. Table 4.2 lists recommended start-up and shut-down procedures for various types of heat exchanger construction.

REFERENCES

1. ASME Boiler and Pressure Vessel Code, An American National Standard.
2. "Standards of Tubular Manufacturers Association," Tubular Exchanger Manufacturers Association, Inc., New York, 6th edition, 1978.
3. Tinker, T., "Shell-Side Characteristics of Shell-and-Tube Heat Exchangers, A Simplified Rating System for Commercial Heat Exchangers," *Trans. ASME*, Vol. 80 (Jan. 1958), pp. 36–52.
4. Palen, J. W. and Taborek, J., "Solution of Shell-Side Flow Pressure Drop and Heat Transfer by Stream Analysis Method" (Heat Transfer Research, Inc., Alhambra, CA), AICHE Chemical Engineering Progress Symposium Series No. 92, Vol. 65 (1969), pp. 53–63.
5. Roark, R. J., "Formulas for Stress and Strain," McGraw-Hill Book Company, Inc., New York, 1965, 4th edition.
6. Timoshenko, S. and Woinowsky-Kreiger, "Theory of Plates and Shells," McGraw-Hill Book Company, Inc., New York, 1959, 2nd edition.
7. Zick, L. P., "Stresses in Large Horizontal Cylindrical Pressure Vessels on Two Saddle Supports," Welding Journal Research Supplement, 1951.
8. Wickman, K. R., Hopper, A. G., and Mershon, J. L., "Local Stresses in Spherical and Cylindrical Shells Due to External Loadings," Welding Research Council Bulletin 107, 1968.
9. ASME "Pressure Vessel and Piping Design," Collected papers 1927–1959, Published 1960.
10. ASME "Pressure Vessel and Piping: Design and Analysis," Vol. 1—Analysis, Vol. 2—Components and Structural Dynamics, 1972.

CHAPTER 5

REFRIGERATION SYSTEMS AND EQUIPMENT IN THE CHEMICAL PROCESS INDUSTRY

RICHARD C. DAVIS
MICHAEL W. OLIVER
Carrier Machinery & Systems Division
Carrier Corp.
Syracuse, NY

INTRODUCTION

The purpose of this section is to discuss refrigeration systems as applied to the chemical process industry. Many chemical processes require a source of cooling in order to remove the heat of chemical reaction, control the temperature of the process, lower the temperature of the process, or change the state of a process fluid.

Often the temperature levels involved are such that ambient air or water from a cooling tower, river, or well can be used.

When the process temperature requirements cannot be met by ambient air or available water sources, a refrigeration system must be employed. The purpose of this section is to discuss the refrigeration systems used to serve the needs of various processes at temperature levels from $-150°F$ to $+70°F$. The types of refrigeration systems discussed in this chapter are:

1. *Refrigeration by Vapor Compression* — using centrifugal, reciprocating and helical rotory "screw" compressors. Emphasis, however, will be placed on the centrifugal compressor systems.
2. *Absorption Refrigeration* — using the water-lithium bromide cycle.

Vapor compression systems are used to:

1. Cool a process liquid
2. Cool water or another secondary coolant
3. Cool a process vapor
4. Condense a process vapor

Absorption refrigeration systems are used primarily to cool process liquids, water or other secondary coolants. However, they can also be used to condense a process vapor.

Other means of producing refrigeration for chemical processes include steam jet systems and expander cycles.

Steam jet systems can be used to cool water down to $40°F$. In addition to water chilling, steam jet systems can be applied to direct vaporization for concentration or dehydration of chemicals. Their use is normally restricted to applications where there are large quantities of steam available in the 25 psig to 125 psig range.

The expander cycle is normally restricted to cryogenic levels (below $-200°$F). Cryogenics are not discussed in this section.

BASIC MECHANICAL REFRIGERATION SYSTEM

The basic mechanical refrigeration system (Figure 5.1) consists of a compressor, condenser, metering device, evaporator and refrigerant. A receiver to store refrigerant is sometimes included in the system — but it is not a mandatory component.

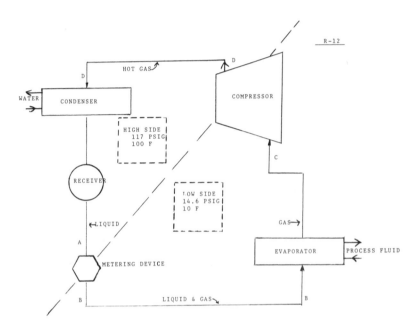

Figure 5.1 Basic refrigeration system.

The compressor is the heart of the refrigeration system. It performs three major functions:

1. It maintains the pressure in the evaporator to permit the refrigerant to vaporize more readily.
2. It compresses low temperature, low pressure vapor from the evaporator to the higher temperature and pressure level of the condenser.
3. It pumps the refrigerant through the system.

The condenser transfers heat from the refrigeration system to a medium that carries it to a final disposal point. Although water is shown as the cooling medium in the condenser, other fluids, including air, are frequently used.

The purpose of the receiver is to hold and store refrigerant. It is generally used only with high pressure refrigerants.

As its name implies, the metering device controls the flow of liquid refrigerant to the evaporator.

The evaporator is the component in which heat is transferred from the process fluid to vaporize the refrigerant.

The refrigeration system operates at two definite pressure levels, the "high side" of the system operates at the saturated condensing pressure and the "low side" at the saturated evaporating pressure.

The "high side" of the system includes all components from the discharge of the compressor to the metering device. The "low side" includes all components from the metering device to the suction side of the compressor. By convention, however, the compressor is regarded as "high side" equipment and the metering device as "low side" equipment.

Another way of illustrating the basic mechanical refrigeration system is by means of a pressure-enthalpy diagram (Figure 5.2).

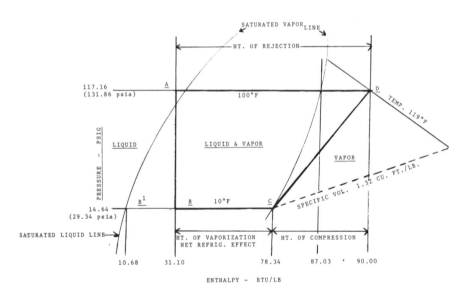

Figure 5.2 Pressure-enthalpy diagram.

The refrigerant leaves the evaporator and enters the compressor as a saturated vapor at Point C. The compression of the refrigerant (Line C-D) raises the temperature and pressure to the condensing level. The enthalpy difference between C and D is the heat of compression. At Point D, the superheated vapor leaves the compressor and enters the condenser. The condensing step (Line D-A) changes the state of the refrigerant from a superheated vapor (Point D) to a saturated liquid (Point A). The heat given up by the refrigerant is carried away by the condensing medium such as water or air. This enthalpy difference between D and A is called the heat of rejection.

At Point A, the hot, high pressure liquid refrigerant enters the metering device. As the refrigerant flows through the metering device, the pressure reduction produces flash gas (Point C) as the remaining liquid is cooled (Point B^1). The resulting enthalpy of the liquid and flash gas mixture is Point B.

At Point B the refrigerant enters the evaporator where the actual cooling takes place. As the refrigerant absorbs heat from the process fluid, the refrigerant changes state to a saturated vapor, Point C. The refrigeration effect is defined as the enthalpy difference between B and C.

The following example will serve to illustrate the use of the pressure-enthalpy diagram.

EXAMPLE

Given: Refrigeration load = 2000 Tons
Evaporator (suction) temperature = $10°F$
Condensing temperature = $100°F$
Refrigerant is R-12

From this information, the following performance calculations can be made from the P-H diagram shown in Figure 5.2.

1. Pressure levels — Evaporator @ $10°F$ = 14.64 psig (29.34 psia)
                  Condenser @ $100°F$ = 117.16 psig (131.86 psia)
2. Refrigeration effect — R.E. = (78.34 − 31.10) = 47.24 BTU/lb
3. Refrigerant flow rate =

$$\frac{2000 \text{ Tons} \times 200 \text{ BTU/min-ton}}{47.24 \text{ BTU/lb}} = 8,467 \text{ lb/min}$$

4. Compressor displacement = 8467 lb/min × 1.32 cu ft/lb = 11,176 cfm
5. Compression ration =

$$\frac{131.86 \text{ psia}}{29.34 \text{ psia}} = 4.5$$

6. Theoretical Compressor BHP: based on isentropic compression (constant entropy)
   a. Heat (work) of compression = (90.0 − 78.34) = 11.66 BTU/lb

   b. Compressor head = 11.66 BTU/lb × 778 $\frac{\text{fr lb}}{B}$ = 9071 ft

   c. Compressor HP = $\dfrac{9071 \text{ ft} \times 8467 \text{ lb/min}}{33,000 \text{ ft lb/min HP}}$ = 2,327 HP

7. Coefficient of performance = $\dfrac{47.24 \text{ BTU/lb}}{11.66 \text{ BTU/lb}}$ = 4.05

8. Compressor discharge temperature = $119°F$

9. Condenser heat of rejection =
   a. Refrigeration effect + heat of compression
      47.24 + 11.66 = 58.9 BTU/lb or
      90.0 − 31.1 = 58.9 BTU/lb
   b. Condenser tons = $\dfrac{58.9 \text{ BTU/lb} \times 8467 \text{ lb/min}}{200 \text{ BTU/min-ton}}$ = 2,494 tons

Actual vs Ideal Cycle

The cycle that was illustrated in the example is a typical ideal cycle. It is based on a number of simplifying assumptions such as:

1. No vapor superheating in the evaporator. Superheating of vapor before compression is normal in many real cycles. The amount varies widely depending upon the installation and type of compressor. Uninsulated suction lines, hermetic motors, txv control sources of suction superheat.
2. No pressure losses except in the metering device. Pressure losses, no matter how small, must occur in every flow process as a necessary part of fluid flow. This can become significant in the suction lines of low temperature systems.
3. No heat transfer except in the evaporator and condenser. Heat flow does occur between the atmosphere and system components. This is usually small but can become significant at low evaporator temperatures or in long runs of piping.
4. No subcooling in the condenser. Some subcooling in the condenser is normal. Very often a subcooler is included in the condenser to improve cycle efficiency.
5. Thermodynamically reversible (isentropic) compression. Actual compression only approaches the ideal isentropic process. The representation of compression at constant entropy is an over-simplification. There are internal energy losses due to friction, windage, and throttling, and, there is heat loss through the casing of the compressor.

For an exact illustration of any cycle, it would be necessary to identify major deviations and to adjust the diagram accordingly.

Despite the fact that it is not precise, the use of the ideal cycle for system representation is valuable. It is quick and easy to use, and does represent the best practical approach to the solution of a complex problem.

HEAT EXCHANGERS

The heat exchangers used in mechanical refrigeration systems for industrial process applications are primarily shell and tube type exchangers. Their use in refrigeration systems are evaporators, refrigerant condensers, and closed economizers.

In some situations, a shell and coil or U-tube type exchanger might also be used. Separate heat transfer coils might also be used in air-cooled condensers and evaporative condensers.

The evaporator (or cooler as it is often called), where the refrigerant changes state from a liquid to a vapor, can be used to:

1. Cool a liquid (i.e. process fluid, water, brine, etc.)
2. Cool a process vapor
3. Liquefy a process vapor

When used with a centrifugal compressor, the evaporator is of the flooded type (Figure 5.3). The refrigerant is on the shell side and the process stream goes through the tubes. This makes cleaning of the process side of the evaporator an easy task. These shell and tube vessels normally include refrigerant distribution system and a vapor separation system.

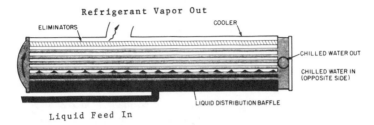

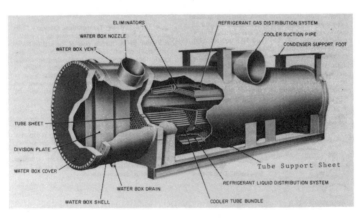

Figures 5.3 Flooded evaporator.

In a reciprocating or screw compressor, the evaporator is normally of the DX (direct expansion) type (Figure 5.4). In this type, the refrigerant goes through the tubes and the process is on the shell side. The use of a DX evaporator simplifies the oil management requirements of reciprocating and screw compressor systems. The process stream must be fairly clean when a DX evaporator is used since the shell side is not so easy to clean.

A flooded evaporator can be used with reciprocating and screw compressor systems, but special attention should be given to continuous return of the oil from the shell side of the evaporator to the compressor.

The refrigerant condenser (Figure 5.5) heat exchanger depends on the method of heat rejection. When water (i.e. cooling tower, river, etc.) is used to condense the refrigerant, a shell and tube heat exchanger is used. The refrigerant condenses on

Figure 5.4 Direct expansion evaporator.

the shell side while the water flows through the tubes. The water side then can be easily cleaned. This is very important due to the water condition of most condenser water systems. Shell and tube type water cooled condensers are usually installed in a close coupled arrangement with the compressor. Therefore, refrigerant piping is minimal.

Air-cooled and evaporative type condensers are also commonly used on recipro- cating and screw compressor systems. With these types, the refrigerant flows through the tubes of a coil and the air (in the case of air-cooled) and air/water spray (in the case of evaporative) flows across the outside surface of the tubes.

Centrifugal systems can also use air-cooled or evaporative condensers. However, there use is normally restricted to areas where water is extremely scarce.

Air-cooled condensers and evaporative condensers are normally located remote from the compressor-evaporator assembly. This results in a more extensive refrig- erant piping system and requires consideration of such parameters as elevation differences, piping pressure drops, additional valves, receivers, etc.

Chart 5.1 provides a relative comparison of the major parameters effecting heat rejection systems. Though this is primarily pointed toward centrifugal compressor systems, it can also be used as a guide for reciprocating and screw compressor systems.

Centrifugal, reciprocating and screw compressor systems can take advantage of subcooling the refrigerant in the condenser to increase the refrigeration effect per pound of refrigerant. Subcooling should not be employed in water cooled con- denser systems where the shutdown equilibrium temperature of the subcooler is below freezing.

Economizers can be used to improve the efficiency of the refrigeration cycles. The flash (open) type and closed type economizers are used primarily on multi- stage centrifugal systems. They can be used on screw compressor systems but cannot be used in reciprocating systems.

The flash economizer (Figure 5.6) employs two stages of flow control between the condenser and evaporator. The flash gas generated as the refrigerant flows through the first control valve is drawn off into an interstage of the compressor. The flash economizer is more efficient than the closed economizer. However, this advantage is minimized at part-load conditions.

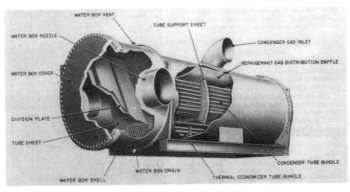

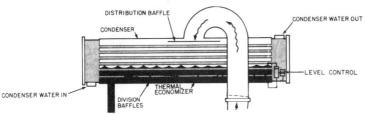

Liquid Refrigerant Out   Refrigerant Vapor from Compressor

Refrigerant Vapor In

Water In and
Out

Refrigerant Liquid Out

Figures 5.5 Water cooled condensers.

RECIPROCATING LIQUID CHILLER

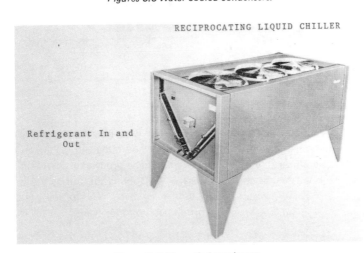

Refrigerant In and
Out

Figure 5.5 Air cooled condenser.

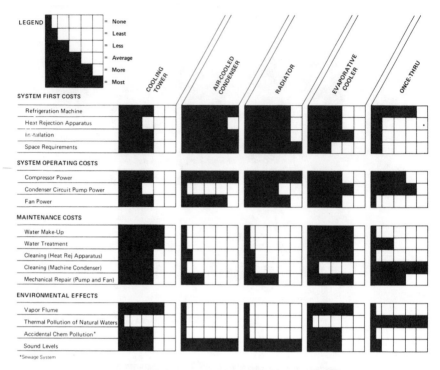

Chart 5.1 Comparison of heat rejection systems.

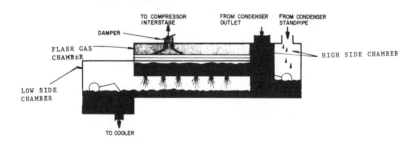

Figure 5.6 Flash economizer.

The closed economizer (Figure 5.7) is a shell and tube heat exchanger which subcools liquid refrigerant on the tube side by evaporating refrigerant on the shell side. Closed economizers are often used in centrifugal process refrigeration systems. The closed economizer has the advantage of subcooling the refrigerant and maintaining the refrigerant pressures near condenser levels. This prevents refrigerant flashing due to piping losses, elevation differences, remote evaporator locations, etc.

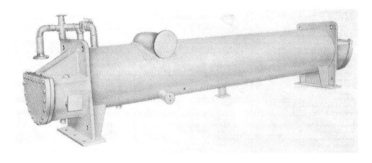

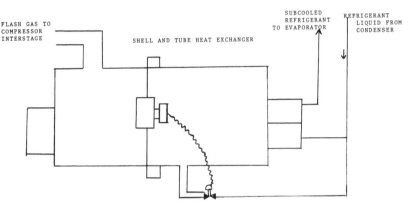

Figure 5.7 Closed economizer.

Other pressure vessels found in refrigeration systems include: receivers, refrigerant storage tanks, suction scrubbers, and oil separators.

REFRIGERANT METERING

The refrigerant metering devices used in refrigeration systems are normally thermal expansion valves (Figure 5.8), for reciprocating and screw compressor systems. Centrifugal systems normally use float operated (Figure 5.9) or pilot operated (Figure 5.10) control valve.

CENTRIFUGAL COMPRESSORS

Centrifugal refrigeration compressors come in a wide variety of shapes and sizes. No two manufacturers build them exactly alike, and no two industrial users have the same specification requirements. With this in mind, the following is a general review of the major components of a centrifugal refrigeration compressor. The specific details may vary with each manufacturer and with each project requirement.

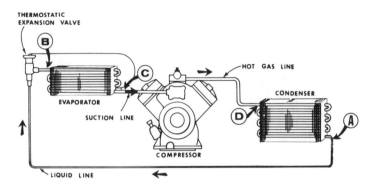

Figure 5.8 Thermostatic expansion valve.

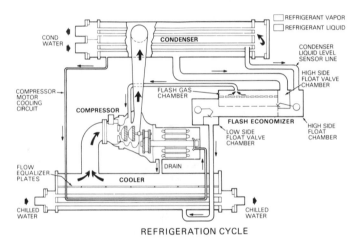

REFRIGERATION CYCLE

NOTE: Float valves in flash economizer meter refrigerant to cooler.

Figure 5.9 Float operated valve.

Centrifugal compressors can be either of the open type or serviceable hermetic type.

Open Drive Compressors

An Open Drive Compressor (Figure 5.11) is one where the driver is outside the refrigerant atmosphere. The compressor shaft extends through the casing. A shaft seal prevents leakage between the refrigerant and ambient air.

Open type centrifugal compressors for process refrigeration are available with one through ten stages and may be driven by most any prime mover. The most common drivers are electric motors and steam turbines. The compressors can often be arranged for drive-through operation. Compressor refrigeration capabilities range

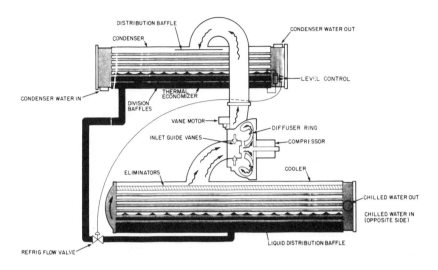

DISTRIBUTION BAFFLE
CONDENSER
CONDENSER WATER OUT
LEVEL CONTROL
CONDENSER WATER IN
THERMAL ECONOMIZER
DIVISION BAFFLES
VANE MOTOR
DIFFUSER RING
INLET GUIDE VANES
COMPRESSOR
ELIMINATORS
COOLER
CHILLED WATER OUT
CHILLED WATER IN (OPPOSITE SIDE)
REFRIG FLOW VALVE
LIQUID DISTRIBUTION BAFFLE

NOTE: Refrigerant flow valve controlled by condenser level control.

Figure 5.10 Pilot operated valve.

2, 3, 4 stages

3 thru 9 stages

Single stage

Figure 5.11 Open drive centrifugal compressors.

85

from 100 tons to 10,000 tons at water chilling levels.

Most centrifugal compressors used in closed cycle refrigeration systems are designed for halocarbon refrigerants such as R-12, R-22, and R-114. Some centrifugal refrigeration compressors can use lighter refrigerants such as ammonia, propane, and propylene.

Often the refrigeration systems which use hydrocarbon refrigerants (i.e. propane, propylene) generally require compressor construction in accordance with American Petroleum Institute Standard API-617, which significantly increases the first cost of the compressor.

The major components of an open centrifugal refrigeration compressor (Figure 5.12) include:

- compressor casing
- shaft
- impellers
- labrynths
- lubrication system

- shaft seal
- journal bearings
- thrust bearing
- capacity control system
- instrumentation

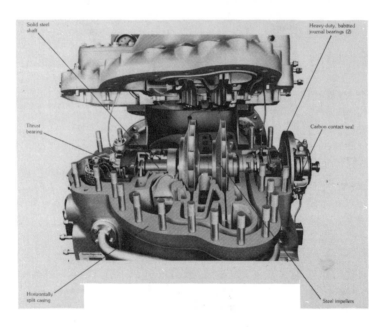

Figure 5.12 Centrifugal compressor cutaway.

Significant attention is usually given to the lubrication system to insure proper lubrication of all rotating parts at all times. In addition to the compressors main shaft driven or electric driven oil pump and lubrication system, a special lubrication package might also be specified (Figure 5.13). This package may also include special

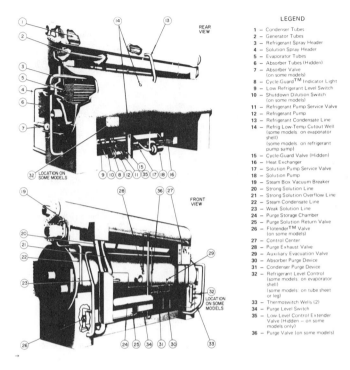

LEGEND

1 — Condenser Tubes
2 — Generator Tubes
3 — Refrigerant Spray Header
4 — Solution Spray Header
5 — Evaporator Tubes
6 — Absorber Tubes (Hidden)
7 — Absorber Valve
      (on some models)
8 — Cycle Guard™ Indicator Light
9 — Low Refrigerant Level Switch
10 — Shutdown Dilution Switch
      (on some models)
11 — Refrigerant Pump Service Valve
12 — Refrigerant Pump
13 — Refrigerant Condensate Line
14 — Refrig Low-Temp Cutout Well
      (some models: on evaporator
      shell)
      (some models: on refrigerant
      pump sump)
15 — Cycle-Guard Valve (Hidden)
16 — Heat Exchanger
17 — Solution Pump Service Valve
18 — Solution Pump
19 — Steam Box Vacuum Breaker
20 — Strong Solution Line
21 — Strong Solution Overflow Line
22 — Steam Condensate Line
23 — Weak Solution Line
24 — Purge Storage Chamber
25 — Purge Solution Return Valve
26 — Flotender™ Valve
      (on some models)
27 — Control Center
28 — Purge Exhaust Valve
29 — Auxiliary Evacuation Valve
30 — Absorber Purge Device
31 — Condenser Purge Device
32 — Refrigerant Level Control
      (some models: on evaporator
      shell)
      (some models: on tube sheet
      or leg)
33 — Thermoswitch Wells (2)
34 — Purge Level Switch
35 — Low-Level Control Extender
      Valve (Hidden — on some
      models only)
36 — Purge Valve (on some models)

Figure 5.13 Auxiliary lubrication system.

instrumentation, valves, piping materials, etc.

Instrumentation can range from the manufacturers standard factory mounted control panel up to a custom designed free standing cabinet (Figure 5.14).

Capacity control of a centrifugal compressor is by variable inlet guide vanes, suction damper, variable speed control, hot gas bypass, or a combination of these.

Part Load Characteristics

For large multi-stage compressors, particularly those with suction damper control, is such that stable performance extends down to approximately 40% of design load, depending on the compressor.

Applications using a 1 or 2-stage compressor with variable inlet guide vanes can modulate effectively down to 10% load or 20 to 35% in a constant lift (head) situation. This is because the guide vanes can effectively modulate 1 and 2 stages of compression but tend to lose their effect in 3 or more stage compressors. Operation below the minimum turndown requires that the cooler either be false loaded or that a hot gas bypass be connected from the condenser directly to the cooler to create enough suction CFM to maintain stable operation of the compressor.

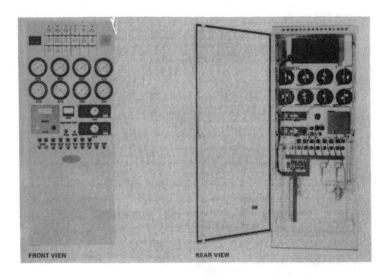

Figure 5.14 Custom control panel.

Where the type of drive permits, variable speed control can be used for part load control. Often speed control would be used in conjunction with one of the other methods to optimize part load efficiency.

Serviceable Hermetic Compressors

A serviceable hermetic compressor (Figure 5.15) is a motor driven compressor with the motor and compressor within the same casing. The motor windings are cooled by the refrigerant in the system. The motor housing can be removed and the motor is accessible for servicing.

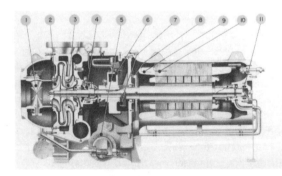

1 — Variable Inlet Guide Vanes
2 — First Stage Impeller
3 — Second State Impeller
4 — Compressor Thrust Bearing
5 — Lubrication Package
6 — Shutdown Oil Reservoir
7 — Dynapoise® Transmission
8 — Centrifugal Demister System
9 — Internal Motor Protector
10 — Internal Motor Protector
11 — Refrigerant Cooling Nozzle

Figure 5.15 Hermetic centrifugal compressor cutaway.

Hermetic motor driven centrifugal compressors are used in many process applications. The hermetic compressor unit is used for cooling process streams, water or mild brine chilling; gas cooling or gas condensing applications.

However, these units are normally available in only one and two stages of compression and, therefore, can only be used for cooling levels down to approximately $10°F$. Hermetic units are available in water chilling level refrigeration capacities of 80 tons to 2000 tons.

The major compressor components are similar to the open compressor except that the hermetic compressor does not require a shaft seal.

Capacity control of hermetic units is by variable inlet guide vane. Hot gas bypass may also be at very low load (below 10%) conditions.

The advantages of the hermetic units include:

1. No shaft seal and related maintenance
2. Motor is not exposed to atmosphere
3. Often has a lower first cost

The hermetic compressor is normally part of a manufacturer's close coupled water cooled packaged refrigeration unit design (Figure 5.16). Cooler, condenser and compressor components can generally be mix-matched to optimize the unit selection.

1 — Hermetic Motor Cooling Line
2 — Flow Equalizer Plates
3 — Refrigerant Distribution System
4 — Condenser Main Tube Bundle
5 — Thermal Economizer
6 — Flash Economizer Float Valve Chamber
7 — Condenser Gas Distribution Baffle
8 — Solid State Control Center
9 — Flash Gas Chamber
10 — Integral Storage Tank
11 — Condenser Float Chamber
12 — Lubrication Package

Figure 5.16 Hermetic centrifugal refrigeration machine.

The hermetic units do provide certain flexibility with respect to customized and special lubrication and instrumentation requirements. These units should be considered where a one or two-stage compressor will meet the required duty.

Application

Centrifugal compressors are used in refrigeration systems serving a wide variety of process applications. Wherever mechanical refrigeration is required in an industrial process, the centrifugal compressor should be considered.

The open centrifugal compressor can be used in various types of cascade systems and drive-through arrangements. Side loads from multiple evaporator levels and economizers can be handled by these compressors. The specific requirements of each process application will govern the most economical and reliable system design.

The capacity of a compressor varies with suction temperature and refrigerant, therefore, it is difficult to discuss capacity ranges of compressors. For example, a given centrifugal compressor using refrigerant R-12 and operating at a condensing temperature of 105°F might have a capacity of 400 tons at −20°F suction temperature while this same compressor would have a capacity of 900 tons at +30°F suction temperature.

Based on a saturated discharge temperature of 105°F, a four-stage open compressor might operate as low as −50°F suction while a ten-stage compressor might operate at −100°F or lower. Refrigerant volume capability of centrifugal compressors ranges from approximately 1000 CFM to 20,000 CFM. For large capacity systems, more than one compressor may be used in the refrigeration system.

The centrifugal compressor regulates to what degree of complexity the system can grow. This is because the number of cycle improvements usually hinge on the number of stages or wheels in the compressor. For example, a single stage compressor has only the option of whether or not to have subcooling whereas a ten-stage compressor could have a primary cooler, two side loads, and six economizers, plus subcooling.

The capabilities of the compressor can also effect the configuration of the refrigerant systems. Side loads will be limited to some degree by the capability of the compressor to handle the associated refrigerant volume. In addition, such considerations as part load performance versus design load optimum break horsepower can effect not only which compressor wheels are used but also the overall design. Examples of refrigeration systems using centrifugal compressors are shown in the Process Cycles Section.

If the refrigeration capacity requirement is small for the required temperature level, a reciprocating or rotary screw compressor might be considered in lieu of the centrifugal compressor. In applications where a single reciprocating or screw compressor will meet the load requirements, there could be a first cost premium if a centrifugal compressor were used. However, other factors evaluated by the process engineer might indicate the preference for a centrifugal compressor. These factors might include: operating cost, horsepower requirements, part load considerations, reliability and maintenance requirements.

RECIPROCATING AND ROTARY SCREW COMPRESSOR SYSTEMS

These compressors are normally used in refrigeration systems where the cooling requirements are too small for a centrifugal compressor. In the smaller tonnages they generally have a lower first cost than centrifugal equipment. However, reciprocating and screw compressors do not have the flexibility for accommodating intermediate level (side) loads and economizer loads. The centrifugal compressor will also often exhibit better part load characteristics.

The reciprocating and screw compressors are used in most all of the industrial process markets.

The duty performed by these compressor systems include: liquid chilling, vapor cooling, vapor condensing and cooling air for a conditioned space.

Some models of these compressors are designed for use with the heavy mole weight halocarbon refrigerants. Others are designed for the lighter mole weight hydrocarbon and ammonia refrigerants.

With reciprocating and screw compressor system designs, special attention must be given to oil management within the system. Since oil flows through the system with the refrigerant, care must be taken to assure that the oil in the system is continuously returned to the compressor.

A complete system includes in addition to the reciprocating or screw compressor:

1. Evaporator: This is normally a direct expansion (DX) air cooling coil or a DX cooler. A flooded type cooler could also be used if required by the application. For ease of oil return to the compressor, a DX cooler would be preferred. However, if the fluid being cooled is highly corrosive or contaminated, a flooded cooler would facilitate cleaning of the process side of the vessel.
2. Condenser: These can be water cooled, air cooled or evaporative cooled.
3. Refrigerant flow control: This is normally a thermal expansion valuve (TXV).
4. Accessories: These include receivers, solenoid valves, strainers, moisture indicators, sight glasses, regulating valves, bypass valves, etc.

Reciprocating

The positive displacement reciprocating compressors are available in open (Figure 5.17), and semi-hermetic (Figure 5.18). The open type has the driver outside the refrigerant atmosphere. The semi-hermetic has the motor inside the refrigerant atmosphere, but it can be removed for servicing. The open compressor can be either direct or belt drive.

These compressors are normally available in sizes ranging from 20 CFM to approximately 600 CFM of displacement. This covers a capacity range of approximately 100 tons of cooling at water chilling levels down to 10 tons at $-50°F$. These capacities are based on $105°F$ condensing temperatures. Manufacturers data should be examined for specific rating information. Larger capacity requirements can be handled by multiple units.

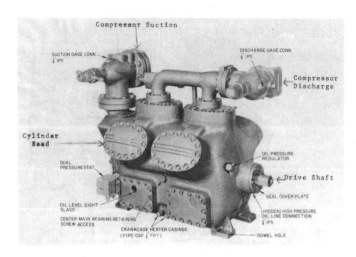

Figure 5.17 Open reciprocating compressor.

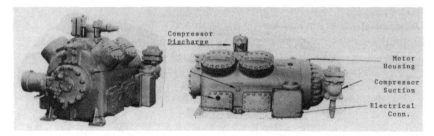

Figure 5.18 Semi-hermetic reciprocating compressor.

The basic reciprocating compressor unit includes the casing, shaft, piston, bearings, shutoff valves, etc., plus a lubrication system and safety devices. Capacity control is by cylinder unloading, hot gas bypass, compressor cycling or a combination. Water cooled heads are sometimes recommended where high discharge temperatures are expected. Water cooled oil coolers are also sometimes required where high suction gas temperatures or excessively high compression ratios are involved.

Figure 5.19 is a schematic of a basic reciprocating system showing the major components. Figure 5.20 shows an assembled package for a multiple compressor system consisting of two refrigerant circuits.

When one stage of compression won't accomplish the required compression ratio, staged or compound systems are employed. A Booster (or low stage) compressor is used to compress the refrigerant from the evaporator pressure up to an intermediate pressure level which is the suction of the high side compressor.

A staged system is essentially a combination of two or more simple system

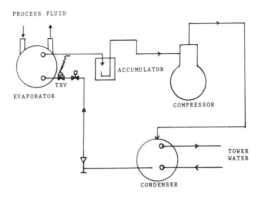

Figure 5.19 Basic reciprocating system.

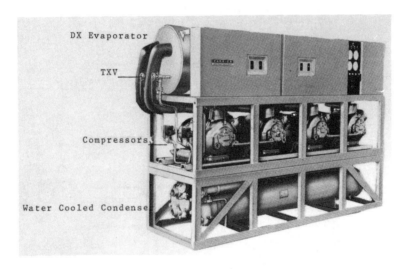

Figure 5.20 Reciprocating chiller package.

cycles. In combining two or more simple flow cycles to form a staged system for low temperature refrigeration, two basic types of combinations are common: direct staging and cascade staging.

Direct staging (Figure 5.21) involves the use of compressors in series, compressing a single refrigerant. The operation of a direct staged system requires cooling (desuperheating) of the gas between stages to avoid excessive heating of the high side compressor. Cooling of the liquid refrigerant supply to the evaporator is also sometimes included in the system.

Cascade staging (Figure 5.22) usually employs two or more refrigerants of

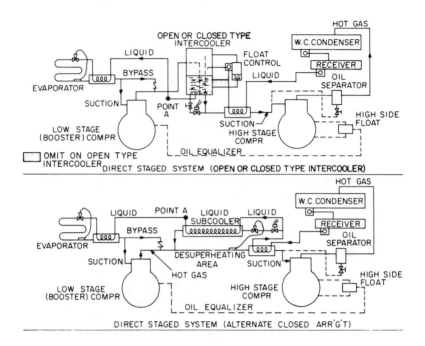

Figure 5.21 Direct staged reciprocating cycle.

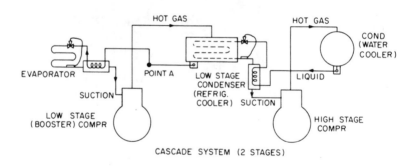

Figure 5.22 Cascade reciprocating cycle.

progressively lower boiling points. The compressed refrigerant of the low stage is condensed in an exchanger (cascade condenser) which is cooled by evaporation of another lower pressure refrigerant in the next higher stage. Figure 5.23 shows a cascade or direct staged industrial reciprocating chiller package.

Screw Compressors

In recent years the helical rotary screw compressor (Figure 5.24) has found

Figure 5.23 Custom packaged reciprocating chiller.
16 tons at −102° F cooling R-11 brine 3-stage cascade using R-503 and R-502; SH120/30 MHP,
SH126/125 MHP; metric controls, special NEMA controls, system includes brine tank, process
and chiller pumps.

Figure 5.24 Screw compressor.

increased usage in the process industries. Initial applications were in the food and beverage industries where the screw compressor systems, using ammonia as the refrigerant, were installed in lieu of reciprocating ammonia refrigeration systems. Today, screw compressor systems are being applied in most all areas of industrial processes when they offer an advantage over centrifugal or reciprocating systems.

As a general guide, the screw compressor normally applies in the tonnage range between the larger reciprocating compressors and small centrifugal compressors. This is where the screw compressor system can be the most competitive. This is especially true where the application calls for compressor ratios (temperature lifts) unattainable with a one or two stage centrifugal compressor and/or above the capacity range of a single reciprocating compressor system. However the screw compressor does overlap the capacities of both the reciprocating and centrifugal compressors.

The compressors are normally available in sizes ranging from 60 CFM to approximately 3300 CFM. This covers a capacity range of approximately 300 tons at $-40°$F up to 1500 tons at water chilling levels.

The operating characteristics of the screw are closely related to the reciprocating machine. The compressor consists basically of these parts: (see Figure 5.25)

1. Low pressure inlet
2. Male driving rotor
3. Female rotor
4. High pressure discharge outlet
5. Slide valve for part load operation

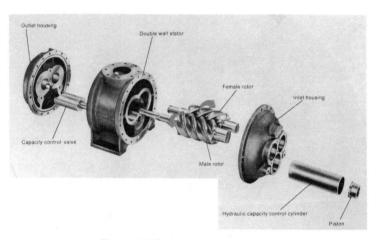

Figure 5.25 Screw compressor internals.

Normally oil is sprayed on the rotors in order to maintain cycle efficiency. The oil acts as a seal and reduces (not eliminates) gas slippage. Some oil is necessarily absorbed by the refrigerant vapor and must be returned to the lubrication cycle. Once the oil and refrigerant are mixed in the compressor, they must be separated, or else the heat exchanger efficiency will be impaired as the lost oil builds up in the refrigeration cycle.

Capacity control is by means of a slide valve which varies the compressor displacement. The slide valve is a movable section of the stator in the rotor bore and allows the gas to return to suction rather than be compressed.

The complete screw compressor assembly normally includes:

1. The compressor with open or hermetic motor drive
2. Slide valve capacity control
3. Lubrication system including oil cooling and oil separator (see Figure 5.26)
4. Control panel
5. Refrigerant shutoff valves, check valves, strainers, etc.

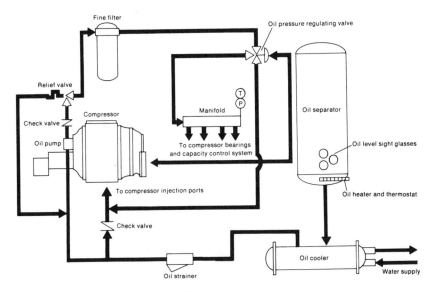

Figure 5.26 Lubrication system.

Figure 5.27 shows the refrigerant flow through the compressor and into the oil separator where most of the oil absorbed in the compressor is removed.

A schematic diagram of a typical screw compressor refrigeration system which includes a closed economizer is shown in Figure 5.28.

The selection of a screw compressor system over a reciprocating or centrifugal is largely one of designer's choice. Items for comparison that have been offered are:

Screw-vs-Reciprocating —

1. Screw compressor has fewer moving parts and therefore potentially less maintenance

2. Screw compressor has smooth capacity reduction rather than stepped increments due to cylinder unloading or compressor cycling
3. Screw compressor can operate down to 10% load
4. Screw compressor is available in larger capacities than reciprocating, thereby eliminating the need to go to duplex or multiple reciprocating compressor systems in those capacity ranges
5. The screw compressor system requires a more complicated oil management system
6. Screw compressor must maintain head pressure to avoid oil foaming problems

Screw-vs-Centrifugal —

1. The choice between a screw compressor system and a centrifugal system is primarily one of cost. In the lower end of the centrifugal compressor capacity range (approximately 1000 CFM) the screw compressor system will normally have a lower first cost
2. Screw compressor must maintain head pressure to avoid oil foaming problems
3. Depending on the specifics of the application, there could be considerations that would still favor a centrifugal system such as horsepower requirements, multiple loads, multiple process temperature levels, etc.

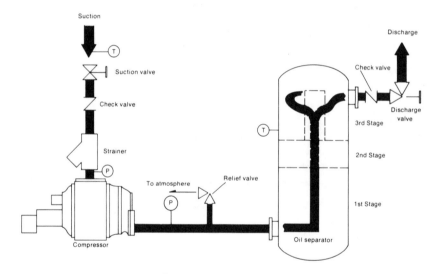

Figure 5.27 Refrigerant flow.

ABSORPTION REFRIGERATION EQUIPMENT

Another method of producing a refrigeration effect is to use an absorption refrigeration machine (Figure 5.29). The absorption machine cycle is based on two principles:

1. Lithium bromide salt solution has the ability to absorb water vapor.
2. Water will boil, or flash cool itself, at low temperatures (i.e. $40°F$) when subjected to a high vacuum or a low absolute pressure (i.e. $0.3''$ hg. abs.)

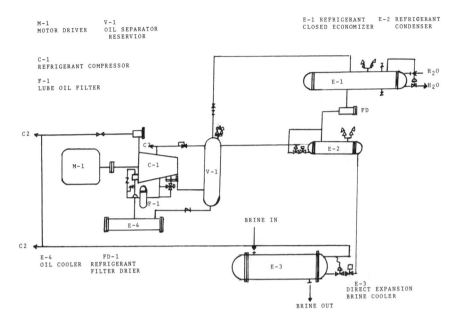

M-1   V-1
MOTOR DRIVER   OIL SEPARATOR
                RESERVIOR

C-1
REFRIGERANT COMPRESSOR

F-1
LUBE OIL FILTER

E-1 REFRIGERANT   E-2 REFRIGERANT
CLOSED ECONOMIZER   CONDENSER

E-4   FD-1
OIL COOLER   REFRIGERANT
              FILTER DRIER

E-3
DIRECT EXPANSION
BRINE COOLER

Figure 5.28 Screw compressor system schematic.

Lithium bromide solution is the absorbent and water is the refrigerant. Heat energy in the form of steam or hot water maintains the concentration of the solution.

The absorption cycle is shown in Figure 5.30. Liquid to be chilled passes through the evaporator tube bundle where it is cooled by the evaporation of the refrigerant water sprayed over the outer surface of the tubes. The refrigerant vapors are drawn from the evaporator to the absorber section where they are absorbed by the solution as it is sprayed over the absorber tubes. The heat which has been removed from the chilled liquid is carried to the absorber by the refrigerant vapors and is transferred to the condensing water flowing through the absorber tubes. The solution in the absorber becomes diluted as it absorbs water, and it is pumped to the generator to be reconcentrated. In the generator, the weak solution is heated by steam or hot water to boil out the refrigerant. The refrigerant vapor passes into the condenser section where it contacts the cool condensing water tubes and condenses. The condensed refrigerant flows back into the evaporator section to start a new cycle. The reconcentrated solution flows from the generator back to the absorber to start a new cycle. Cycle efficiency is improved by passing the relatively cool weak solution from the absorber and the relatively warm strong solution from the generator through a heat exchanger.

A basic equilibrium diagram for lithium bromide in solution with water is shown

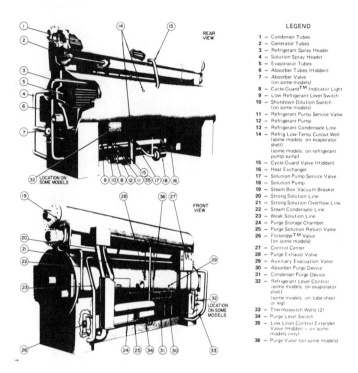

LEGEND

1 — Condenser Tubes
2 — Generator Tubes
3 — Refrigerant Spray Header
4 — Solution Spray Header
5 — Evaporator Tubes
6 — Absorber Tubes (Hidden)
7 — Absorber Valve
   (on some models)
8 — Cycle-Guard™ Indicator Light
9 — Low Refrigerant Level Switch
10 — Shutdown Dilution Switch
   (on some models)
11 — Refrigerant Pump Service Valve
12 — Refrigerant Pump
13 — Refrigerant Condensate Line
14 — Refrig Low-Temp Cutout Well
   (some models on evaporator
   shell)
   (some models, on refrigerant
   pump sump)
15 — Cycle-Guard Valve (Hidden)
16 — Heat Exchanger
17 — Solution Pump Service Valve
18 — Solution Pump
19 — Steam Box Vacuum Breaker
20 — Strong Solution Line
21 — Strong Solution Overflow Line
22 — Steam Condensate Line
23 — Weak Solution Line
24 — Purge Storage Chamber
25 — Purge Solution Return Valve
26 — Flotender™ Valve
   (on some models)
27 — Control Center
28 — Purge Exhaust Valve
29 — Auxiliary Evacuation Valve
30 — Absorber Purge Device
31 — Condenser Purge Device
32 — Refrigerant Level Control
   (some models, on evaporator
   shell)
   (some models, on tube sheet
   or leg)
33 — Thermoswitch Wells (2)
34 — Purge Level Switch
35 — Low Level Control Extender
   Valve (Hidden — on some
   models only)
36 — Purge Valve (on some models)

Figure 5.29 Absorption refrigeration machine.

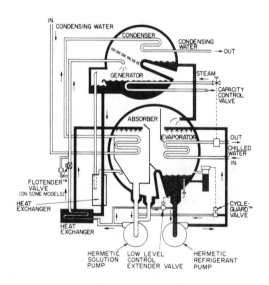

Figure 5.30 Absorption cycle diagram.

in Figure 5.31. Points 1 through 7 represent a complete absorption refrigeration cycle. These values will vary with different loads and operating conditions.

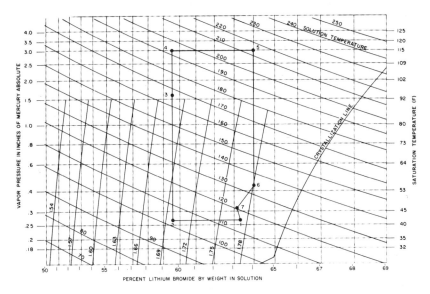

Figure 5.31 Equilibrium diagram.

An explanation of each point and lines drawn between is as follows:

Point 1 represents the strong solution as it begins to absorb water vapor in the absorber.
Point 2 represents the weak solution leaving the absorber and entering the heat exchanger after having absorbed water vapor.
Point 3 represents the weak solution leaving the heat exchanger with the same concentration as at Point 2, but at a higher temperature.
Point 4 represents the weak solution in the generator after being preheated to the boiling temperature.
Point 5 represents the maximum solution concentration in the generator after the refrigerant has been boiled out.
Point 6 represents the strong solution leaving the heat exchanger after losing heat to the weak solution.
Point 7 represents strong solution leaving the spray nozzles.

The absorber, evaporator and condenser are shell and tube vessels which normally use copper or cupro-nickel tubes. However, other tube materials could be used. The generator is normally either shell and tube type or U-tube using cupronickel tubes.

The absorption machine also includes a purge unit (mechanical or thermal) to remove any air or noncondensibles that leak into the machine. Controlling the heat flow to the generator controls the strength of the reconcentrated solution, and hence, the ability of the solution to absorb refrigerant water vapor and cool the process fluid.

The capacity control valve modulates steam or hot water to the generator and is positioned by a controller which senses leaving chilled fluid temperature. At full load conditions the control valve is wide open. As the load is reduced and the chilled fluid temperature drops below design temperature, the control valve will be throttled. At no-load conditions the control valve will be closed.

The absorption machine is used in industrial process primarily for liquid chilling duties. Leaving fluid temperatures as low as 40°F can be produced. The refrigerant (water) which freezes at 32°F, prevents operation at lower temperatures. Heat is removed from the machine through the water cooled condenser.

The energy input to the machine is by low pressure steam, (15 psig or lower) or hot water (or other liquid) in the 200°F to 300°F range.

The energy input requirements of an absorption machine is approximately 18 lbs of steam per hour, per ton of refrigeration. Nominal capacities available range from 70 tons to 1200 tons.

A double effect type absorption machine uses high pressure steam (i.e. 150 psig). The energy requirement of this type of machine is in the area of 12 lb/hr per ton of refrigeration.

The absorption unit's use of hot water in the generator often becomes involved in many energy studies using waste heat to produce chilled water for cooling or process needs. Solar applications using solar panels to generate hot water can also be used. The steam requirements of the absorption machine can be met using waste steam, discharge of back pressure turbines, etc.

The major advantage the absorption unit, provided there is a need for the chilled water, is the use of alternate fuel sources. Toward this end, many firms rely on absorption equipment to reduce their dependence on one utility.

PROCESS REFRIGERATION CYCLES

The following discussion of industrial process centrifugal refrigeration cycles will provide insight regarding the advantages and disadvantages of the various types of systems. Each process application will require different considerations since each product and plant have different requirements for first cost, operating cost, control stability and ease of operation.

Open Cycle

An open cycle refrigeration system is basically one in which all, or part, of the process refrigerant gas that has been compressed never returns to the compressor. This type of system is typical of liquid storage applications, such as ammonia, propane, butane, etc. Figure 5.32 shows a basic open cycle refrigeration system. In this system, the process refrigerant vapor from the storage tank passes through a

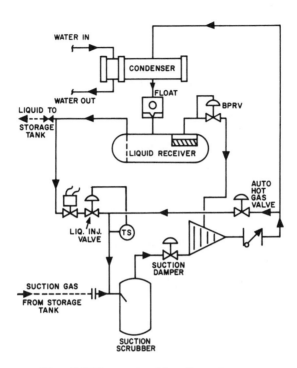

Figure 5.32 Open cycle refrigeration system.

suction scrubber, a device to prevent liquid droplets from entering the compressor, and enters the compressor where it is compressed to sufficient pressure to be condensed by a cooling media, which, in this case, is water. The condensed refrigerant liquid then passes into a high pressure float valve. As the liquid passes through the float valve, flashing of a portion of the liquid occurs. This is due to a reduction of pressure as it enters the liquid receiver, and at essentially compressor interstage pressure, since the liquid receiver is vented to an intermediate compressor stage.

The flash gas which was formed passes into the remaining compressor stages and is re-compressed. The remaining saturated liquid, which has been cooled by the flash gas, passes out of the liquid receiver to the storage tank. A refrigerant pump is not normally required to transfer liquid from the receiver to the storage tank since sufficient pressure differential normally exists to use the refrigerant pressure in the receiver as the driving force to transfer the liquid.

Not shown in Figure 5.32 is a refrigerant sub-cooler (i.e. closed economizer) which is sometimes required. A closed economizer is used in cases where the saturated refrigerant liquid would flash before reaching the expansion valve at the storage tank. By sufficiently sub-cooling the liquid refrigerant, piping pressure losses and liquid "static head" losses will not cause flashing of the liquid as the sub-cooled refrigerant liquid will never drop below saturation temperature and pressure.

Another consideration for an open system of this type is the amount of fluctuation of the process load or flow requirements. Figure 5.32 shows a centrifugal compressor in the system (reciprocating and rotary screw compressors are also commonly used depending on load and temperature levels). Centrifugal compressors are capable of handling high process flow rates, such as those experienced during periods when the storage tank is being filled or loaded. However, when the storage tank is not being loaded, the only real source of generated vapor is heat leakage from the outside ambient into the tank. Insulation of the tank normally keeps the "holding" refrigeration load to about 5–10% of the "loading" situation. Since horsepower requirements of the centrifugal compressor may not drop below roughly 30% of design, due to the requirement of discharge gas bypass to maintain stable operation, the centrifugal compressor becomes very inefficient at greatly reduced loads. For this reason, a reciprocating, or other relatively small capacity compressor system, is also often furnished with the system to handle the small flow requirements experienced during a "holding" period. The system brake horsepower is then reduced by more fully loading the smaller compressor.

Closed Cycle

A closed cycle refrigeration system is one in which the entire refrigerant charge is contained in a closed loop and is continuously compressed, condensed and vaporized within the system. There are as many different closed cycle refrigeration system designs as there are processes requiring refrigeration. In addition, for a given process refrigeration requirement, there are usually several viable system approaches which will accomplish the job. The system design approach will be influenced by basic factors such as initial system cost, operating costs, and owning costs. Of course, there are also many factors which influence component selection and system design. These include: required system capacity, process evaporator temperature level(s), condenser cooling media (water, air, etc.), system power consumption, physical space limitations, noise limitations, etc.

Figure 5.33 shows a basic closed cycle refrigeration system and the P-H diagram of a cycle for a cooling process with a single evaporating temperature using a multi-stage centrifugal compressor.

The heat removed from the process fluid is absorbed by the refrigerant (shell side) and causes the refrigerant liquid to vaporize. The refrigerant vapor enters the 2-stage compressor and is compressed up to a pressure (condensing pressure) that will allow the vapor to be condensed by cooling water (tube side). The vapor condenses on the shell side of the condenser and drains into a single stage flash economizer.

The purpose of the economizer is to reduce the compressor brake horsepower by reducing the mass flow of refrigerant required at the process evaporator. The liquid

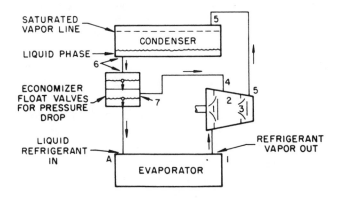

2-Stage Compression Cycle
with Economizer

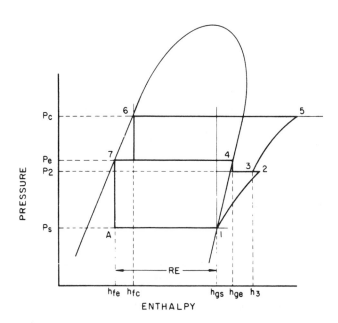

Pressure-Enthalpy Diagram, 2-Stage
Compression Cycle with Economizer

Figures 5.33 Closed cycle refrigeration system.

refrigerant enters the flash economizer where some refrigerant is flashed and re-compressed while the remaining liquid refrigerant is fed to the cooler. This completes the closed cycle as the liquid, which enters the cooler (evaporator), again vaporizes due to heat gain from the process fluid.

Level operated control valves may sometimes be used in lieu of the float operated valves. The cycle shown in Figure 5.33, when used with a 4-stage compressor, and Refrigerant R-12, is capable of operating as low as $-40°F$ to $-50°F$ suction temperatures and is used for a wide range of capacity requirements. The efficiency of this cycle can be improved using a two-stage economizer as shown in the P-H diagram Figure 5.34. The additional economizer increases the refrigeration effect of the liquid refrigerant entering the cooler and decreases the refrigerant flow through the first stage of the compressor. This same cooling requirement can be handled in a number of different ways. One arrangement is to combine the economizer into a storage tank. Another arrangement is the use of multiple evaporators if the brine rise were high, i.e. 20 to 40 degrees.

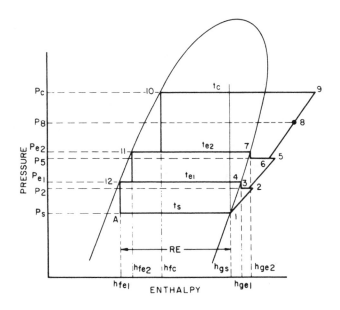

Figure 5.34 P-H diagram — 2 stage economizer system.

In such a case, it would be possible to use two evaporators in series, each taking roughly half the load. The higher level would be vented to the second stage inlet, the lower to the first stage inlet. Often times, this arrangement can prove to be more efficient than the use of only one heat exchanger.

With this cycle, it is also possible to have the entire system factory packaged or shipped in separated components. Single or multiple coolers may be involved. The cycle is, therefore, quite flexible from an application standpoint. Refrigerants 114, 500, and 22 can also be used with this cycle. The refrigerant flow diagram for this system is shown in Figure 5.35.

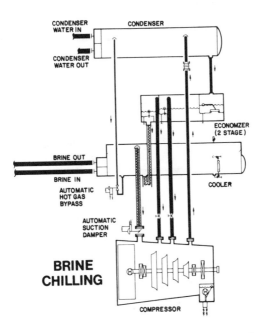

Figure 5.35 System flow diagram — 2 stage economizer.

Remote Evaporator

A common method of dealing with a situation where the process heat exchangers are located remotely from the refrigeration machine is to cool a brine and then pump it to the heat exchangers.

Sometimes this situation can be handled by adding a closed economizer to the refrigeration cycle. This subcools the high pressure condensed refrigerant, enabling it to be transferred by its own pressure to a remotely located cooler or liquifier, thereby eliminating the brine system.

The cycle in Figure 5.36 is typical of those in which the process load is a significant distance away from the compressor. The vapor to be condensed is supplied to the tube side of the vessel. The refrigerant is on the shell side where it is monitored by a level control through a refrigerant feed valve. There may be a single evaporator (as shown) or multiple evaporators, all of which would be under level control. One of the design considerations in this particular system is the availability

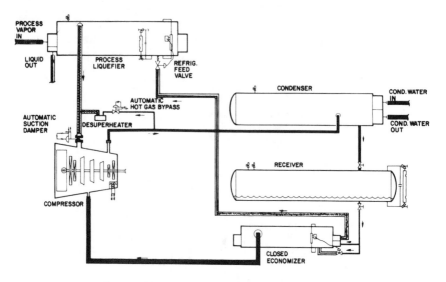

Figure 5.36 Remote evaporator refrigeration system.

of refrigerant at adequate pressure to feed the remote liquefier (evaporator).

Note that the condensed refrigerant leaves the condenser and flows into an in-line receiver which is maintained under full condenser pressure. Normally, this receiver is also the refrigerant storage tank and is sized to contain the entire charge of the system. It is also an operating receiver, in that it provides a space in the system for the changes in refrigerant level which result from changes in load.

From the operating receiver the liquid is fed to the tube side of a closed economizer. This heat exchanger subcools the liquid (still under pressure) to a temperature approaching the compressor interstage pressure. Subcooling takes place at part of the liquid refrigerant and is expanded across a feed valve (controlled by a level controller) into the shell side of this vessel which has vapor connection to the suction of the compressor second stage wheel. Therefore, the liquid is subcooled, allowing it to be fed to a remote location under full high-side pressure without flashing in the liquid line. Excessive flashing in the liquid line can produce very erratic action at the refrigerant feed valve and should be avoided. The P-H diagram is shown in Figure 5.37.

Control of this system is maintained by an automatic suction damper and automatic hot gas bypass. Because of the remote liquefier, the automatic hot gas bypass is taken from the discharge line of the compressor back into the suction line through a de-superheater. The de-superheater is a device which injects liquid refrigerant into the hot gas bypass in order to remove the superheat.

Figure 5.38 depicts a slightly more complicated system which also makes use of a single multi-stage compressor. The P-H diagram for this system is shown in Figure

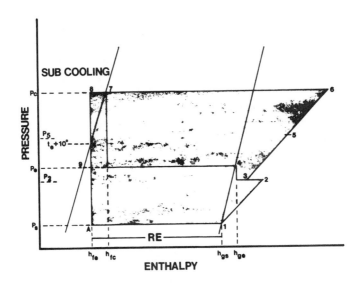

Figure 5.37 P-H diagram — remote evaporator.

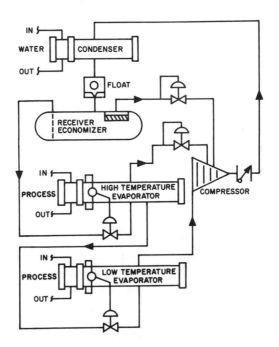

Figure 5.38 Multiple evaporators — series.

109

5.39. This system includes an open economizer, a side load (high temperature evaporator), and a low temperature evaporator. The side load cooler provides an economizer effect in feeding the primary cooler as it pre-cools the entering liquid refrigerant to the primary cooler. The use of a side load to provide a flash (open) economizer effect can only be accomplished if the side load and the primary load vary directly. This is common on many processes since the loads vary with the amount of product being processed. In some cases, the side load is not related to the primary load. In these cases it is necessary to feed both the side load and the primary load from the same source to prevent refrigerant feed problems (see Figures 5.40 and 5.41).

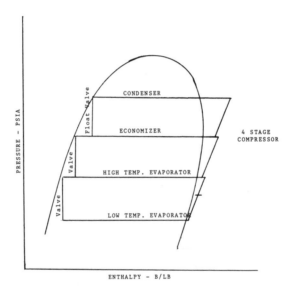

Figure 5.39 P-H diagram — series evaporators.

The side load allows the designer of a multi-stage system to accommodate cooling loads of different temperatures using a single refrigeration system. Another application of the side load is on a single cooling load involving a high temperature rise of the process stream. By using a side load, the process stream can be piped, in series, first through the side load cooler and then through the primary cooler. Here, the side load acts as an open economizer and reduces the BHP requirement below that of a single cooler on the same process stream.

Often, the specific capacities and configurations of a manufacturer's compressor will dictate cycle or system alterations. With side loads, for instance, the maximum CFM allowable is dependent on compressor design. If the side load CFM is larger than the maximum, sometimes a blank stage is introduced into the wheel line-up to

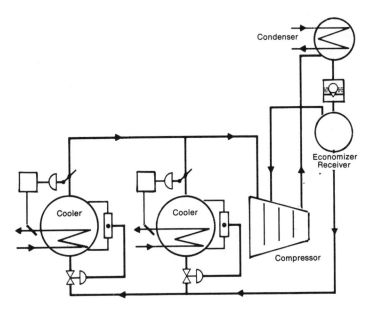

Figure 5.40 Multiple evaporators — parallel.

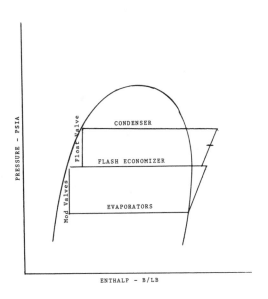

Figure 5.41 P-H diagram — parallel evaporators.

reduce the inlet velocities to the wheel. Another limitation is the lift capability of the compressor.

To overcome lift limitations or to handle a large side-load, a drive-thru arrangement is used. The drive-thru connects the compressor, in series, to a single driver. On the refrigerant side, the discharge of the first stage is connected directly to the suction of the second. Often a large side load is also connected to the suction of the second compressor. The single and drive-thru compressor systems are quite versatile in handling varied system demands, an illustration being the application of this kind of system to vapor condensing such as ammonia condensing.

Ammonia Condensing

Figure 5.42 is a typical system used for condensing a vapor such as ammonia in the tubes of the cooler. As the ammonia vapor is not compressed, it remains oil-free. The condensed liquid is available to the system with its full latent heat. Since the refrigerant in the cooler is colder than the ammonia, the ammonia vapor from the receiver migrates to the cooler where it is condensed, the liquid ammonia returning, by gravity, to the receiver.

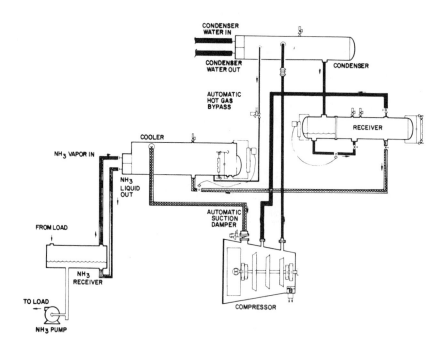

Figure 5.42 Ammonia condensing system.

The ammonia loads are served by pumping the liquid ammonia from the receiver to the loads where it is vaporized and returned to the receiver. Assuming that the system uses R-114, R-12, R-22, or R-500, a pump out system and storage tank are required for servicing the refrigeration equipment. As shown, the storage tank can also be utilized as an economizer. A separate chamber is built into the end of the receiver to act as a condensed liquid sump. A liquid level controller maintains the appropriate level in this chamber while refrigerant is fed through a feed valve into the economizer-receiver portion of the vessel. A vapor connection is taken off the top and piped to the suction of the second stage wheel in the compressor. The liquid in the economizer-receiver is then available to feed the cooler through another refrigerant feed valve controlled by a level controller on the cooler. The refrigerant cycle for this three stage compressor and single stage economizer is shown in Figure 5.43. The flash gas is introduced between the first and second stages of the compressor.

The advantages of an ammonia condensing system are:

1. The ammonia remains oil-free. Since the ammonia does not pass through the compressor, it does not pick up any oil to contaminate the system. As a result, the ammonia cycle does not lose efficiency. The need for oil traps is also eliminated.
2. The system is easy to operate. Pumping the ammonia involves no more equipment than pumping any other type of brine.
3. Relatively constant system pressure. This means that all valves in the ammonia system can be selected for the same pressure.
4. The ammonia has a higher latent heat at colder temperatures. The higher latent heat at the colder supply temperature will reduce the amount of ammonia supplied.
5. Smaller liquid and vapor lines are required, as fewer pounds of ammonia are circulated.

Of course, there are some disadvantages in the ammonia condensing system.

1. Insulation required on liquid line. Ammonia boils at $-28°$F. If the liquid lines were not insulated, the ammonia would vaporize in the piping instead of in the process heat exchangers.
2. An ammonia receiver is required. In a straight compression system, liquid ammonia would be pumped directly to the process heat exchangers. A separate vessel would not be required as it would be in this case.

Staged Refrigeration

Staged refrigeration systems, unlike the single compressor systems previously described, involve two independent compressor drive trains, and to a certain extent, two combined refrigeration systems.

Staged refrigeration should be considered when:

1. The temperature levels involved require compressor lifts in excess of the capabilities of a single compressor.
2. Higher system efficiency is required (lower system BHP).
3. The cooling requirements include loads at two or more temperature levels, especially when the load at the higher temperature is quite large relative to the load at the lowest temperature level.

113

4. Smooth and efficient system response is required particularly in side load applications. Since staged systems size the compressors to a particular load (high or low side), fewer compromises have to be made in their selection. This offers the potential of better turndown, and higher efficiency at part load conditions.

Staged refrigeration systems can be used to produce cooling levels to as low as $-135°$F. The two types of staged refrigeration are:

1. Direct staged refrigeration.
2. Cascade refrigeration.

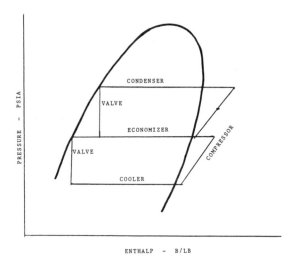

Figure 5.43 P-H diagram ammonia condensing.

Direct Staged Refrigeration

The system shown in Figure 5.44 requires the use of the same refrigerant in the high and low stages. Although two temperature levels of brine can be obtained, this is essentially a seven stage machine. Refrigerant from the three stages of the low side compressor discharges directly into the suction of the four-stage high side compressor through a desuperheater.

Some desuperheating is normally required to avoid overheating the high side compressor. The high side compressor then discharges into the condenser. The cycle illustrated in the flow schematic shows a single stage open economizer. The leaving liquid refrigerant feeds the high stage cooler. The low stage cooler is fed with liquid from the high stage cooler. Note that a surge drum is inserted in the line to insure a space for vapor and liquid separation, and to provide a source of solid liquid to feed the automatic control valve on the low stage cooler. Without this surge drum, refrigerant vapor may be carried into the liquid feed line causing erratic action of

114

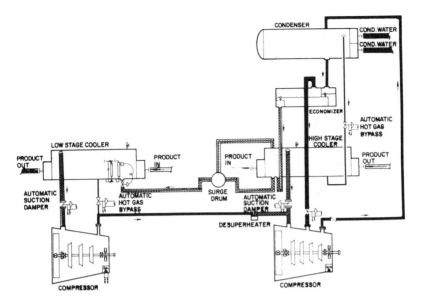

Figure 5.44 Direct staged refrigeration.

the liquid feed valve. The two compressors on this system can be optimized separately to provide minimum horsepower.

This system, however, is more difficult to operate than other types of staged systems. Also, if there is any oil leakage within the system, the oil will collect in the low stage cooler where it will be the most troublesome. Figure 5.45 is the P-H diagram for the direct staged cycle just described. Note the clear cut separation of the low and high sides.

The advantages of a direct staged refrigerant cycle are:

1. It takes less power to operate than a comparable cascade system and, often, comparable single compressor or drive-thru systems.
2. The cycle uses only one refrigerant.
3. A single pumpout, storage tank and purge are required as only one refrigerant is used.
4. The system is less expensive than a cascade system.
5. It can provide two levels of cooling and accommodate large cooling loads at the higher temperature level.
6. The loads at the two temperature levels can vary, independently.

The disadvantages of a direct staged refrigeration cycle are:

1. The system is more difficult to operate than a cascade system.
2. The high and low side loads must match the compressor capabilities.

Basic Cascade System

Figure 5.46 is a flow diagram of a basic cascade refrigeration system. The

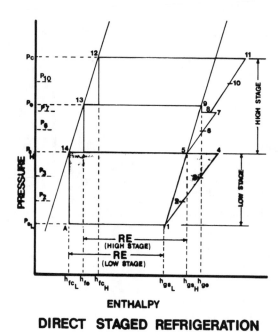

DIRECT STAGED REFRIGERATION

Figure 5.45 P-H diagram direct staged refrigeration.

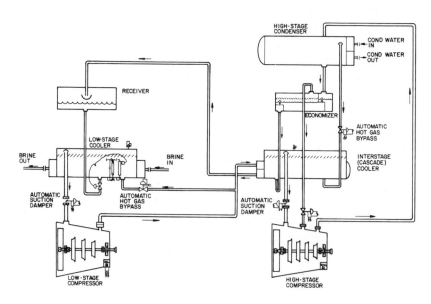

Figure 5.46 Basic cascade system.

operating simplicity and reliability of this cycle make it suitable for single process temperature applications as low as −135°F.

The cascade system differs from the direct staged system in that the cascade system has two separate refrigerant circuits, thereby allowing the use of different refrigerants on the high and low stages. The direct staged system, of course, uses one continuous refrigerant.

The heart of the cascade refrigeration system is the interstage or cascade cooler. The cascade cooler serves as the cooler of the high stage cycle and as the condenser of the low stage cycle.

Different refrigerants can be used in each cycle to provide a close match-up with the process load requirements and optimize system efficiencies. Figure 5.47 is the P-H diagram for the basic cascade system.

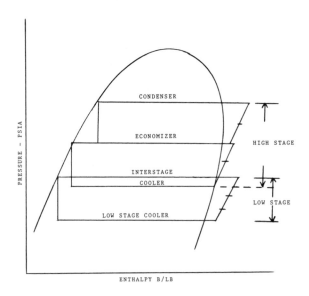

Figure 5.47 P-H diagram basic cascade.

The system would be used where the process temperature level is below the capability of a single compressor system. Instead of a drive-thru or single compressor system, a cascade system can also be used when reduced compressor horsepower is required.

Cascade System — Refrigerant Side

The cascade-refrigerant side system (Figure 5.48) has all of the elements of the basic cascade system just discussed, including two separate refrigerant circuits.

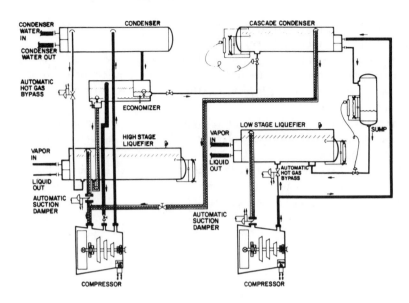

Figure 5.48 Cascade system — refrigerant side.

(Low stage refrigerant is condensed in a vessel that also serves as a cooler for the high stage cycle.) Essentially, it is a combination of the basic cascade and direct staged refrigeration systems. The system is different, however, in that a high stage cooler is also included. This cooler permits the system to accommodate cooling loads at two different temperature levels. The process load can be either a liquid or a vapor. In order to condense the low side refrigerant, the low side condensing temperature must be greater than the high side evaporator temperature.

It is also possible to apply such a system with reciprocating compression equipment in either the high or low stage where loads indicate such equipment choice desirable.

For example, with a very small low stage load — typical of chlorine tail gas recovery, amounting to four to five tons of refrigeration at $-80°F$ — a reciprocating cycle may be indicated.

The high stage evaporator on such a system would normally have a large load and require a centrifugal. The reciprocating compressor of the low stage can be cascaded in the manner shown into the high stage machine without any direct refrigerant connection.

Cascade systems may cost slightly more, and require slightly more horsepower than direct staged systems, but they are often easily justified because of their operating simplicity, flexibility, reliability and stability.

The P-H diagram in Figure 5.49 shows the overlap between the low and high stages.

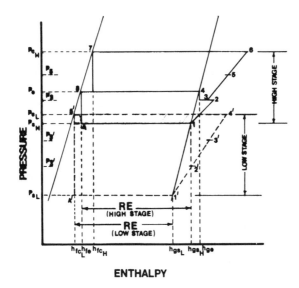

ENTHALPY

Figure 5.49 P-H diagram refrigerant side cascade.

The advantages of the Cascade System are:

1. Different refrigerants can be used in each stage for optimum compressor efficiency.
2. One low side cycle can be used with several high side cycles.
3. The system is easy to operate.
4. The condensing medium for the low side cycle can be brine from the high side.

The disadvantages of this refrigeration cycle are:

1. It is more expensive to operate.
2. It requires more power than a direct staged system.

Whether or not these are really disadvantages must be tempered by the fact that more work can be accomplished. Moreover, this system does provide cooling at two temperature levels.

Cascade System — Brine Side

Figure 5.50 represents another variation of a staged refrigeration cascade system. Instead of linking both machines on the refrigerant side, they are linked by the secondary coolant, i.e. brine. The brine from the high stage machine is used as the cooling medium in the condenser of the low stage machine.

This is also a variation of the basic cascade system. The only difference between this system and the basic cascade system is that brine, in lieu of refrigerant, is used as the cooling in the low stage condenser.

This cycle is particularly useful when there are refrigeration loads at two different temperature levels. For example, there might be load requirements for $10°F$

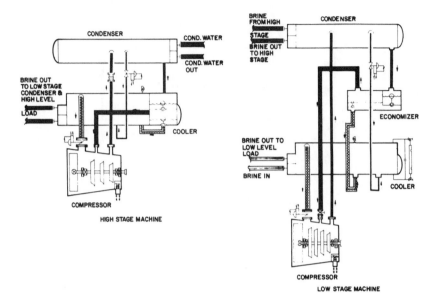

Figure 5.50 Cascade system — brine side.

and −90°F brine. The left side of Figure 5.50 illustrates a close coupled brine machine which is used to produce the 10°F brine.

As an additional load on the high stage machine, however, some of the 0°F brine is circulated to the condenser of the low stage machine, shown on the right side of Figure 5.50. The low side condenser heat of rejection is added to the original 0°F brine requirement. The low stage machine would operate at approximately −100°F suction and +20°F condensing.

The system is an extremely simple system to operate. It is particularly useful where different brines are required in the high and low stages, as well as providing the flexibility of using different refrigerants in the high and low stages.

SYSTEM DESIGN CONSIDERATIONS

All refrigeration systems require proper application. Provided below are some considerations in system and component requirements.

Liquid Slugging

Almost all open cycle systems should have a suction knockout drum or suction scrubber to prevent liquid carryover or liquid "slugging" to the compressor. Closed cycle systems with separately purchased evaporators often require knockout drums since the liquid eliminators are not always designed for refrigeration duty by those heat exchanger manufacturers outside the refrigeration industry. Also, on a system with a remote, elevated evaporator it is possible, during down periods, to condense refrigerant in the vertical suction pipe from the evaporator to the compressor. This usually occurs when the ambient temperature is low — such as during winter or on a cool night. When liquid is formed in the vertical suction line, it will collect in the low spot of the piping. It is important, therefore, to design suction piping with no traps or U-bends. If liquid is allowed to collect, and possibly fill a suction line pocket, the compressor can be damaged on start-up when the liquid "slug" quickly passes into the compressor. The knockout drum will prevent most of the liquid from reaching the compressor. The same analysis holds true for compressor lines from interstage economizers.

Flashing Upstream of Control Valve

The liquid feed valve (usually controlled by liquid level) at the chiller must be supplied with 100% liquid. A common system mis-design is to fail to provide 100% liquid to the liquid feed valve. This is usually encountered on remote process evaporator applications (often with an elevated evaporator). Saturated liquid from the condenser will flash with any significant pressure drop due to either hydraulic friction losses in the piping and valves or pressure loss from a vertical "liquid head" of refrigerant. If the evaporator is mounted 40 feet above the saturated liquid refrigerant source (condenser, receiver, etc.), for R-12 at 110°F, the liquid static head (neglecting friction losses) is:

$$40 \text{ ft} \times 77.38 \#/\text{ft}^3 /144 \text{ in.}^2 /\text{ft}^2 = 21.5 \text{ psi}$$

To prevent this flashing, adequate pressure must be provided to reach the valve and adequate subcooling to insure the liquid does not flash before it reaches the valve.

Elevated Evaporator Drainback

On systems with elevated, flooded evaporators, Figure 5.51, particular attention should be given to the "drain-back" of refrigerant to a lower elevation. On a large tonnage process chiller, there could be several thousand pounds of refrigerant on the shell side. On system shutdown, the liquid level feed valve (on liquid level control) will open, allowing refrigerant liquid to drain out of the chiller and back to a lower elevation. It is therefore, very important to provide a refrigerant liquid receiver sized to hold 100% of the system charge. If a liquid receiver is not provided, the entire charge in the evaporator could drain out and conceivably fill the condenser and piping to the compressor. The end result is compressor filled with liquid and potential compressor damage if the casing is not drained prior to start-up.

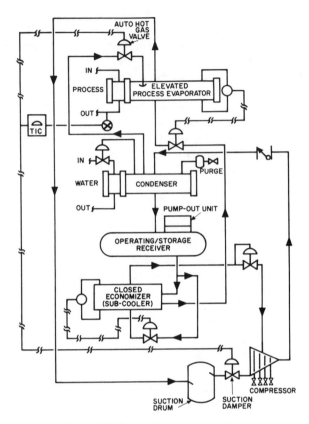

Figure 5.51 Elevated process evaporator.

Piping Pressure Drops

Refrigeration pressure drops throughout the refrigeration circuit must be analyzed, but the most critical pressure drop is in the main compressor suction line. For example, if the actual pressure drop in the suction line from the evaporator to the compressor inlet flange is 1 psi higher than the allowed pressure drop, this will result in a 4°F higher evaporator temperature on R-12 at −40°F. It will also cause the leaving process temperature to rise approximately 4°F, further causing the refrigeration unit to fail to perform as specified. Moreover, particular attention should be paid to any system components which are mounted a considerable distance from the compressor or refrigeration system. Refrigeration lines should *never* be sized by the velocity method under such circumstances.

Torque Capability of Driver

The torque required by the refrigeration compressor to accelerate from 0% to 100% speed must be considered, particularly on low temperature applications where torque requirements are the greatest. If the compressor ever requires more torque than the driver can deliver, the compressor will stall out (stop accelerating) and fail to achieve design speed. Electric motor, steam turbine with limited steam supply, and gas turbine (single shaft) are types of drives which require particular attention. Such drives should be checked against the breakaway torque, starting torque and the full load torque requirements of the compressor. Depending on the type of drive and the system environment of the driver, each of the torque requirements may create difficulties. The compressor should be set up to accelerate to speed within 5 to 15 seconds on an electric motor, depending on starter type.

Rapid System Pressure Equalization

Immediately after a system has been shut down, the pressure existing on the high side of the system attempts to equalize with the low side pressure by whatever refrigerant flow paths which exist. A slow pressure equalization is desirable for a number of reasons:

   a. To prevent drive train reverse rotation (compressor-gear-driver), which may cause mechanical damage or, indirectly, damage by loss of proper lubrication.
   b. On a water-cooled condensing application, with evaporator temperature below 20°F, the risk of freezing water in the condenser tubes (if there is a sub-cooler circuit in the condenser) and the possible rupturing a tube becomes a problem. Such problems increases with lower evaporator temperatures. The system, therefore, must be designed to slow equalization by such methods as check valves, fast closing suction and interstage valves, and auto hot gas valves, the latter being particularly important.

Refrigerant Migration

Refrigeration migration is an often overlooked phenomenon experienced during system down-time periods. In a system which has just been shutdown, the temperature and pressure of the system will start to equalize and reach equilibrium. On low temperature applications in particular, the compressor suction end casing and rotor assembly, which have large masses heat up at a slower rate than the bulk of the refrigerant in the system. Occasionally, the still cold surfaces of the compressor and casing will actually condense refrigerant vapor, with the liquid accumulating in, or migrating to, the bottom of the compressor casing. Compressor damage can also occur if the compressor is restarted without draining the casing. Casing stage drains are to be used to "blow down" the casing, removing any liquid that might have accumulated.

System Optimization

To illustrate how these different refrigeration systems can be used to satisfy the same process load, consider the following problem: Cool a process stream from +70°F to −10°F with cooling water available at 90°F. Evaporator load = 400 tons (4.8 million BTH).

Solution #1 (Figure 5.52) − This is a "bare bones" low cost refrigeration system which will provide the necessary cooling capacity. Note the required compressor BHP.

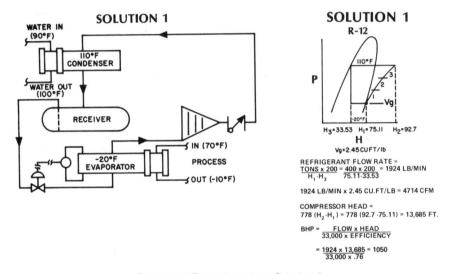

SOLUTION 1

WATER IN (90°F)

110°F CONDENSER

WATER OUT (100°F)

RECEIVER

−20°F EVAPORATOR

IN (70°F)

PROCESS

OUT (−10°F)

SOLUTION 1

R-12

110°F

P

3
2
1 — Vg

−20°F

$H_3 = 33.53$ $H_1 = 75.11$ $H_2 = 92.7$

H

$Vg = 2.45 CUFT/lb$

REFRIGERANT FLOW RATE =
$$\frac{TONS \times 200}{H_1 - H_3} = \frac{400 \times 200}{75.11 - 33.53} = 1924 \, LB/MIN$$

1924 LB/MIN × 2.45 CU.FT/LB = 4714 CFM

COMPRESSOR HEAD =
778 ($H_2 - H_1$) = 778 (92.7 − 75.11) = 13,685 FT.

$$BHP = \frac{FLOW \times HEAD}{33,000 \times EFFICIENCY}$$

$$= \frac{1924 \times 13,685}{33,000 \times .76} = 1050$$

Figure 5.52 Example problem Solution 1.

Solution #2 (Figure 5.53) − This solution shows how compressor BHP can be reduced by the addition of one stage of economizing. Note the reduction in compressor BHP.

Solution #3 (Figure 5.54) − This shows how BHP can be reduced further by chilling the process stream in two steps. The high temperature evaporator will perform ½ of the total cooling by chilling the process stream from 70°F to 30°F. The remainder of the load will be handled by a separate low temperature evaporator, which cools the process stream from +30°F to −10°F. Note the savings in compressor BHP.

Summary (Table 5.1) − While Solution #1 has the lowest first cost, Solution #3 might be the best overall choice. Depending on energy costs, the payback period for Solution #3 could be as short as one year. While the specific system first costs may vary, the relative difference can serve as a guide when comparing higher first cost to reduced energy consumption.

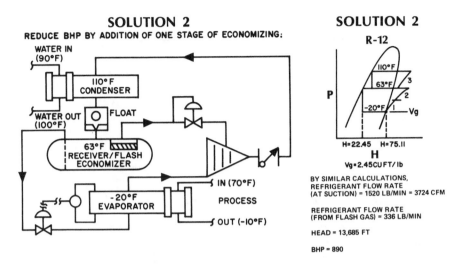

Figure 5.53 Example problem Solution 2.

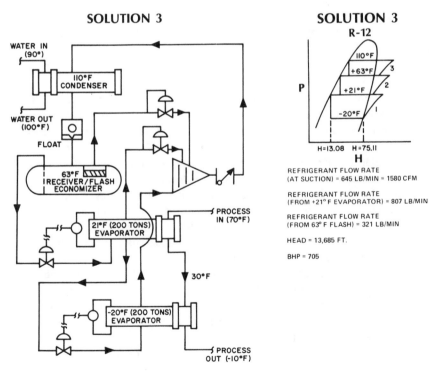

Figure 5.54 Example problem Solution 3.

Table 5.1. Problem Summary.

Solution	# 1	# 2	# 3
BHP (TOTAL)	1050	890	705
%POWER REDUCTION	0%	15%	33%
SUCTION CFM	4714	3724	1580
RELATIVE COMPRESSOR SIZE	LARGE	MEDIUM	SMALL
RELATIVE SYSTEM FIRST COST INDEX	1.0	1.2	1.4

REFRIGERATION ECONOMICS

The design of a refrigeration system and selection of its components quite often requires some economic decision making. The example in the previous section provided the basis for such decisions. Other parameters to be examined and alternatives to choose from include:

1. Absorption versus mechanical refrigeration.
2. High energy consumption versus larger heat transfer surfaces.
3. Water cooled versus air cooled or evaporative condensing.
4. Evaporator temperature level versus its affect on the process.
5. Lowest first cost versus lowest energy consumption.

Mechanical Refrigeration

The approximate energy input requirement for a compressor in a mechanical refrigeration system can be seen in Figure 5.55 as a function of the compressor suction temperature. This is typical for centrifugal, reciprocating and screw compressor systems. Exact values will vary with individual compressor, refrigerant and cycle efficiencies. At the lower suction temperatures, direct staging or cascading of compressors can reduce the horsepower to the lower portions of the curve.

Note, also the effect of condensing temperature on horsepower requirements. A water cooled or evaporative cooled condenser system where ambient wet bulb temperature is 78°F might typically have a 105°F condensing temperature. An air cooled system where the ambient dry bulb temperature is 95°F might have a condensing temperature of 120°F or higher.

The screw and reciprocating compressors will probably require a booster stage when operating below −20°F suction with a 120°F condensing temperature.

Part load horsepower characteristics will depend on the specific type of compressor devices, method of capacity control and the relationship of condenser media temperature to part load.

The horsepower requirement of a compressor can be calculated from:

$$BHP = \frac{\text{WEIGHT FLOW (LB/MIN)} \times \text{HEAD (FT)}}{\text{COMPRESSOR EFFICIENCY} \times 3300 \text{ (FT LB/MIN-HP)}}$$

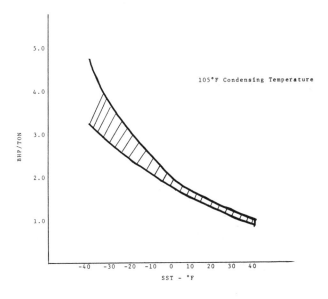

Figure 5.55 BHP/TON vs SST curve.

For a given operating point or range, the manufacturer will select the most efficient impeller(s) from those available.

The refrigerant weight flow can be reduced by adding cycle efficiency improvements such as subcooling as flash economizers. Depending on the type of compressor and flexibility of the total refrigeration machine design, these improvements may or may not be possible.

The most flexibility the designer has over the factors affecting capacity is the temperature lift (head).

The designer can decrease the compressor horsepower by raising the suction temperature and/or lowering the condensing temperature. This will reduce the total work of the compressor required to "lift" the refrigerant from suction to condensing conditions. If the temperature of the condensing medium is as low as possible at design conditions and the evaporator temperature level has been fixed by the process, then one way to reduce the lift is to increase the amount of heat transfer surface in the evaporator and/or condenser. In the equation: $Q = U.A.LMTD$, for a given cooling load [Q], if the area (A) is increased, then the log mean temperature difference (LMTD) can be reduced. This results in lower temperature lift (i.e. higher suction temperature and/or lower condensing temperature) and, therefore, reduced compressor horsepower.

Figure 5.56 shows the effect of leaving temperature difference (LTD) on required heat transfer surface for several secondary coolants. The designer must evaluate the savings in compressor horsepower compared to the cost of additional heat transfer surface.

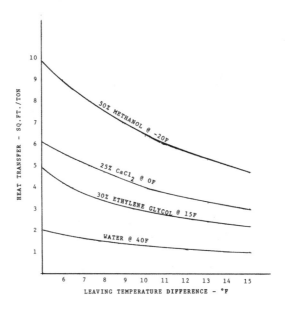

Figure 5.56 LMTD vs sq ft surface.

The temperature lift can also be reduced by increasing the heat transfer co-efficient (U) through the use of some of the newer high performance heat transfer surfaces.

Absorption Refrigeration

The energy requirement of an absorption machine is essentially that energy required in the generator to boil off the refrigerant water from the weak lithium bromide solution and thereby reconcentrate the solution. The electric driven solution and refrigerant pumps require minimal (½ HP to 7½ HP) energy.

The steam consumption rate for an absorption machine being supplied with 14 psig steam is approximately 18 lb/hr per ton of refrigeration. This steam rate can increase to 21 lb/hr per ton as steam pressure drops to a minimum of (0) psig. Condenser water requirements increase significantly as steam pressure drops. Nominal capacities of most absorption machines are based on 12 to 14 psig steam.

A hot fluid (hot water or other process fluid) in the 200°F to 300°F range can also be used to operate an absorption machine. To realize full capacity from the machine, the fluid temperature should be in the 250°F to 300°F range. The absorption machine will require approximately 17 MBH of hot fluid per ton of refrigeration.

The double effect absorption machine, using two generators, operates on steam at 150 psig or lower and has a steam rate of approximately 12 lb/hr per ton of refrigeration. While the first cost of the double effect machine is substantially higher it does offer a lower steam rate where the higher pressure steam is available.

Comparison

The choice of absorption versus mechanical refrigeration is normally based on several factors, including:

1. Available energy sources.
2. Cost of various energy sources.
3. Required refrigeration capacity and process temperature level.
4. Available condenser heat rejection means and temperature level.
5. Space requirements.
6. Process requirements.
7. Owner and operator preference and familiarity.
8. Relative cost of auxiliaries (pumps, fans, water treatment, etc.)

Figure 5.57 shows an operating cost comparison of absorption versus mechanical refrigeration as a function of energy cost. This data is for the refrigeration machine only and is based on a 40°F leaving process temperature and 85°F condenser water and 10 to 14 psig steam or equivalent.

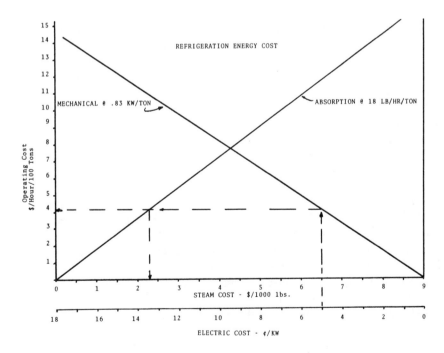

Figure 5.57 Energy cost: Absorption vs mechanical.

The designer must evaluate these factors as they apply to each specific project.

WASTE HEAT RECOVERY

In the past, energy recovery of low level heat sources has not been widely attempted. The additional large, and sometimes expensive, equipment required and the cheap, easily available energy made most recovery systems economically unjustifiable. Today, with the soaring cost of energy, systems which optimize the use of energy or which convert waste energy to useful purposes can often be justified and applied.

The use of refrigeration equipment can often provide an efficient and economic means of waste heat recovery. Refrigerant use is attractive because of the moderate pressure levels, reasonable cost, thermal stability, non-flammability, non-toxicity and non-corrosiveness with common metals.

Attention must be given to proper refrigerant selection due to the instability of some fluorocarbons at elevated temperatures. The pressure-temperature relationship should also be examined to avoid excessive pressures at the higher temperatures. The use of a refrigerant such as R-114 provides a stable, medium pressure, fluorocarbon with which to work.

Industrial Heat Pump

A reasonably standard refrigeration system can be applied to high temperature heat pump systems. The successful application of this system depends on a fairly constant year round heat sink plus a requirement for the higher temperature level. Typical applications include chemical processes, sewage treatment plants and nuclear power plants.

Figure 5.58 is the basic high temperature heat reclaim cycle which is recovering waste heat from a process water stream. This waste heat would normally be rejected (i.e., through a cooling tower). The reclaimed heat is pumped through the compressor up to the higher temperature level of the refrigerant condenser. Here the waste heat is transferred to the hot water system where the heat can now be put to beneficial use for process loads or space heating.

A coefficient of performance (COP) of 3 to 4 can be obtained using available low level (100° to 180°F) heat sources and raising them to higher, more usable levels. Figure 5.59 shows the approximate COP and BHP per ton as a function of compressor lift. (Condensing temperature minus suction temperature.) This is based on using R-114 and an economizer in the system.

Compressor lift can be estimated by assuming a 10°F leaving temperature difference in the evaporator and condenser.

Referring to Figure 5.58, assume 480 GPM of waste process water at 105°F.

$$\text{Evaporator Load} = \frac{480 \text{ GPM} \times (105 - 85)}{24} = 400 \text{ tons}$$

Desired leaving condenser water temperature $= 180°F$
Temperature lift $= (180 + 10) - (85 - 10) = 115°F$
From Figure 5.59 — Estimated COP $= 3.5$

Compressor BHP/ton $= 1.9$
Compressor BHP $= 1.9 \times 400 = 760$ BHP

Available condenser heat $=$ Evaporator load $+$ compressor work

$$= 400 \text{ tons} + .212 \frac{\text{Tons}}{\text{BHP}} \times 760 \text{ BHP}$$

$$= 561 \text{ tons condenser heat.}$$

This means that 561 tons of heat (6732 MBH) at $180°F$ can be produced for process duty or space heating.

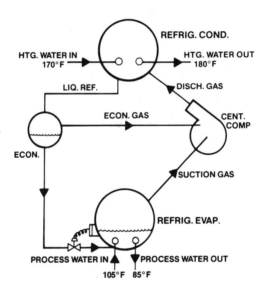

Figure 5.58 Industrial heat pump cycle.

Power Generation

Figure 5.60 shows the use of a refrigeration machine for a rankine cycle application to produce useful shaft horsepower. This system, which uses the compressor as an expander can be used to drive a generator to produce electricity or the shaft horsepower can drive most any mechanical equipment.

Referring to Figure 5.60, assume a steam load to the evaporator or boiler of 105,000 lbs/hr at 12 psig and a total latent heat of evaporation of approximately 100 million BTUs/hr. Table 5.2 summarizes the significant data.

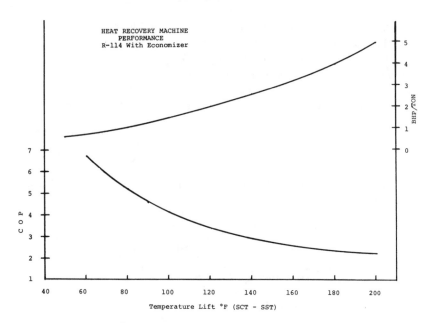

Figure 5.59 Lift vs COP and BHP/TON curve.

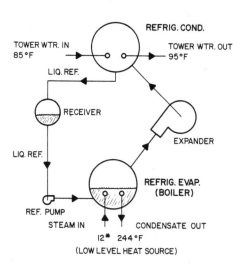

Figure 5.60 Power generation cycle.

There are many more variables that need to be analyzed and accounted for. However, this example should provide some insight as to the potential for the cost and energy savings through this system.

Table 5.2. Rankine Cycle.

- 105,000 lb/hr. steam to evaporate - 12 psig
- 100 million BUT/hr. - 8333 tons
- 85°F condenser water
- Feed pump - 550 HP, other losses 50 HP
- Net shaft HP = 4000 HP
- Generator @ 95% Eff. - 2800 KW
- Cycle Eff. = $\frac{Output}{Input}$ = 9.5%

Steam Generation

Another example of the possible use of refrigeration machines for energy recovery is steam generation. Where the temperature level of the waste heat is too low for waste heat boilers, the refrigeration cycle shown in Figure 5.61 may prove to be economical.

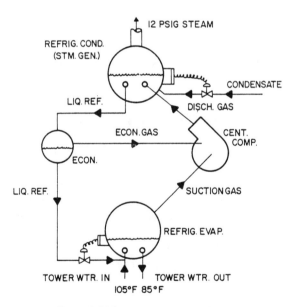

Figure 5.61 Steam generation cycle.

An attempt has been made to show various methods of applying refrigeration machinery to waste heat recovery applications involving "low level" heat sources. It is believed that as power costs continue to increase, utilization of low level

133

heat sources cannot be neglected. The refrigeration machine as a recovery device will prove to be an economical method to realize these economies.

REFRIGERANTS

A refrigerant is any fluid that picks up heat by evaporating at a low temperature and pressure and gives up heat by condensing at a higher temperature and pressure.

Theoretically, there are few limitations on the number of substances that can be used in refrigeration systems. Practical considerations such as thermodynamic properties, toxicity and chemical stability have limited the commonly used refrigerants. The refrigerants used today include halocarbons such as: R-12, R-22, R-114, and R-500. Ammonia and hydrocarbons such as propane and propylene are frequently used in process refrigeration applications to meet specific requirements.

No substance is the "ideal" refrigerant. A refrigerant is selected only after the characteristics of available refrigerants are matched against the requirements of the system. Generally, the selection will be a compromise between conflicting desirable properties. In many cases, the choice will also be governed by properties that are not directly related to the refrigerant's capacity to remove heat. For example, flammability, toxicity and availability will often be the deciding factors in the selection of a refrigerant.

The important thermodynamic properties are: pressure, temperature, volume, and enthalpy.

The important physical properties are: oil solubility, leakage tendency, odor, toxicity, flammability, moisture reaction and leak detection.

Thermodynamic and physical properties for every refrigerant are published in various sources.

The pressure-temperature relationship of some common refrigerants is shown in Figure 5.62.

The effects of various thermodynamic and physical properties are summarized in Table 5.3.

Halocarbon, hydrocarbon and ammonia refrigerants are all used for process refrigeration systems. Refrigerant selection should include an analysis of physical and thermal properties of the refrigerants as they apply to each project. The impact of refrigerant selection on equipment cost, operating cost, auxiliaries and code requirements should all be analyzed.

Table 5.4 is a tabulation of comparative data for various refrigerants used in process refrigeration systems.

Heat Transfer Fluids

In many refrigeration applications, process heat is transferred, not to the primary refrigerant used in the vapor compression cycle, but to a secondary coolant, which may be "any liquid cooled by the refrigerant and used for the

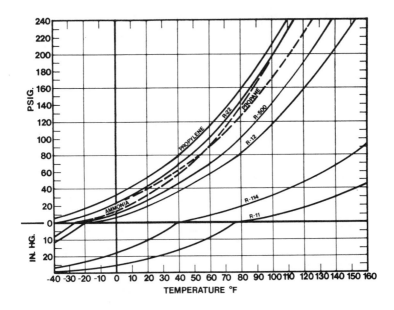

Figure 5.62 Pressure temperature curve.

Table 5.3. Properties vs System Effect.

REFRIGERANT PROPERTY	EFFECT ON SYSTEM
PRESSURE TEMPERATURE VOLUME DENSITY	DETERMINE SIZE, STRENGTH AND TYPE OF COMPONENTS.
ENTHALPY	QUANTITY OF REFRIGERANT IN SYSTEM AND ITS RATE OF CIRCULATION.
PHYSICAL PROPERTIES SUCH AS : TOXICITY, ODOR, FLAMMABILITY, LEAKAGE RATE, ETC.	NO EFFECT ON HEAT TRANSFER ABILITY. INFLUENCE CHOICE OF EQUIPMENT.

Table 5.4. Comparative Data.

	12	22	114	290	500	502	717	1270
NAME	Dichlorodi-fluoromethane	Monochlorodi-fluoromethane	Dichlorotetra-fluoromethane	Propane	Azeotrope R-12/152a	Azeotrope R-22/15	Ammonia	Propylene
Chemical Formula	CCL$_2$F$_2$	CHCLF$_2$	C$_2$CL$_2$F$_2$	C$_3$H$_8$	73.8% CCl$_2$F$_2$ 26.2% CH$_3$CHF$_2$	48.8% CCl$_2$F$_2$ 51.2% CClF$_2$CF$_3$	NH$_3$	C$_3$H$_6$
Molecular Weight	120.93	86.48	170.93	44.10	99.29	111.64	17.03	42.09
Boiling Temp at 1 ATM (F)	−21.6	−41.4	38.4	−43.7	−28.0	−50.1	−28	−53.9
Freezing Temp at 1 ATM (F)	−252	−256	−137	−305.8	−254	*	−107.9	−301
Critical Temp (F)	233.6	204.8	294.3	206.3	221.1	194.1	271.4	197.2
Critical Pressure (Psia)	597	716	474	617.4	631	618.7	1657	670.3
Density Liquid @ 100 F	78.79	71.24	88.4	29.58	69.28	71.97	36.4	30.3
Specific Vol. Vapor @ 0 F	1.61	1.37	4.75	2.68	1.66	0.88	9.12	2.26
Specific Heat Liquid @ 100 F	.240	.313	.249	.6727	.306	.308	1.158	.609
Liquid Head Ft/Psi @ 100 F	1.84	2.04	1.65	4.89	2.10	1.98	3.96	4.74
Saturation Pressure (Psia)								
At: −40 F	9.31	15.22	1.91	16.09	10.95	18.8	10.41	20.59
0 F	23.85	38.66	5.95	38.34	27.98	45.78	30.42	47.95
20 F	35.74	57.73	9.69	55.76	41.96	67.16	48.21	69.16
100 F	131.86	210.60	45.85	188.25	155.90	230.89	211.90	227.58
125 F	183.77	292.64	67.55	257.18	217.7	316.06	307.8	308.97
Thermal Conductivity (K)								
Sat. Liquid − 0 F	.0490	.0630	.0437	.0680	.0530	.0469	.3350	.0780
Sat. Vapor − 100 F	.0060	.0068	*	.0126	*	.0071	.0180	.0116
Viscosity − Centipoise								
Sat. Liquid − 0 F	.3133	.2670	.5994	.1575	.2823	.2728	.2282	.1253
Sat. Vapor − 100 F	.0132	.0140	.0121	.0091	.0130	.0142	.0117	.0096
Basic Cycle: 0 F/100 F								
Refrigeration Effect B/Lb	46.2	65.2	38.1	108.2	55.8	40.1	457	115.4
Liq. Circulated Lb/Min-Ton	4.33	3.07	5.25	1.85	3.58	4.98	.438	1.73
Volume Flow Cfm/Ton	6.97	4.20	24.93	4.95	5.95	4.38	3.99	3.93
C.O.P.	3.66	3.50	3.62	3.42	3.50	3.26	3.62	3.43
Safety Group − U.L. Class	6	5a	6	5b	5a	5a	2	5a
Safety Group − ANSI B9.1	1	1	1	3	1	1	2	3
Explosive Range (% by Vol.)	Nonflammable	Nonflammable	Nonflammable	2.3−7.3	Nonflammable	Nonflammable	16−25	2.0−10
Cost − Compared to R-12	1.0	1.58	1.40	0.22	1.49	2.49	.14	.94

*Not available.

transmission of heat without a change in its state." These secondary coolants are sometimes known as heat transfer fluids, brines, or secondary refrigerants.

In the process heat exchanger, the "brine" extracts heat from the process and transports it to the evaporator where the heat is given up the refrigerant. Brine, then, is the vehicle by which sensible heat is transferred from the process to the refrigerant.

Some heat transfer fluids require special heat exchanger tube materials. Table 5.5 is a guide to recommended material selections.

Table 5.6 shows properties and relative cost of some commonly used secondary coolants.

SUMMARY

It has been the intent of this section to provide an overview of various types of mechanical and absorption refrigeration systems used in the chemical process industries. A discussion of the components making up those systems with emphasis on the compressors was also included. This should provide the chemical engineer with general guidelines regarding the parameters affecting refrigeration systems and the types of equipment available.

The Charts shown in Tables 5.7 and 5.8 provide further insight as to the variables affecting refrigeration system design, the data that should be made

Table 5.5. Tube Materials.

CONTACT-SURFACE METAL	COPPER	CUPRO-NICKEL 90-10	CUPRO-NICKEL 70-30	RED BRASS 85	ADMIRALTY BRASS	ALUMINUM BRASS	STEEL SAE 1010	NICKEL STEEL 3½%	STAINLESS STEEL 304L	TITANIUM
RELATIVE COST (1)	1.00	1.65	2.17	1.73	2.00	1.74	1.04	2.65	6.03	4.44
SODIUM CHLORIDE CALCIUM CHLORIDE (2)	1	4		2	3			5		
METHYLENE CHLORIDE (3)	2			3				1		
TRI-CHLOROETHYLENE & METHANOL (3)	2							1		
ETHYLENE & PROPYLENE GLYCOLS (3)	1	2	3					4	Note (5)	Note (5)
FLUORINATED REFRIGERANTS (3)	1							2		
WATER FRESH	1	3	6	2	5	4				
WATER BRACKISH	1	4	3	2	2					
WATER SEA		2	1		4	3				
Vapor Condensing Processes										
AMMONIA CHLORINE (DRY) (4)							1	2		
ETHYLENE										

NOTES:
1. Always check current prices for specific applications.
2. The corrosion resistances of copper and red brass in contact with these brines are similar when the proper amount of brine inhibitor is used. For this reason, the table lists copper as the economical first choice instead of red brass, even though red brass is used much more frequently.
3. Nickel steel is preferred over copper on low temperature applications whenever the shell material is nickel steel. Tube and shell materials should be identical to avoid differential thermal expansion stress.
4. Nickel steel used in preference to ordinary steel because of an improved Charpy impact value at low temperature.
5. Stainless steel and titanium are shown only to give relative cost. High cost will normally rule out their use except to meet customer specifications.

Table 5.6. Common Properties.

Heat Transfer Fluids commonly used with Carrier Refrigeration Machines. This table is for use as a guide in selecting Heat Transfer Fluids for process applications.

TEMP	HEAT TRANSFER FLUID (F)	% SOL./WT	SPEC. HEAT BTU/#	SPEC. GRAV.	VISC CENT.	THERM. COND. BTU/HR/SQ. FT	GPM/TON/10 F	hb *	FREEZ. PT (F)	BOIL. PT (F)	SOL. $/GAL.
+30	Sodium Chloride	12	.86	1.093	2.2	.28	2.55	941	17.5	215.0	$0.02
	Calcium Chloride	12	.83	1.109	2.4	.32	2.62	971	19.0	213.0	0.05
	Methanol Water	15	1.00	0.986	3.2	.28	2.45	781	13.5	187.0	0.22
	Propylene Glycol	30	.94	1.034	8.0	.26	2.47	240	13.0	216.0	0.71
	Ethylene Glycol	25	.92	1.037	3.7	.30	2.52	775	12.9	217.0	0.73
	Methylene Chloride	100	.275	1.362	0.54	.097	6.45	662	−142.0	103.6	3.18
	Trichloroethylene	100	.223	1.500	0.74	.068	7.22	473	−126.0	185.0	3.38
	Refrigerant 11	100	.201	1.537	0.55	.069	7.82	517	−168.0	74.8	5.13
+15	Sodium Chloride	21	.80	1.167	4.2	.25	2.57	693	1.0	216.0	0.03
	Calcium Chloride	20	.72	1.199	4.8	.31	2.77	730	1.0	214.0	0.08
	Methanol Water	22	.97	0.968	5.3	.26	2.56	599	4.5	182.0	0.32
	Propylene Glycol	40	.89	1.050	20.0	.24	2.58	67	−4.2	218.0	0.98
	Ethylene Glycol	35	.86	1.058	6.8	.28	2.65	576	0.0	219.0	1.02
	Methylene Chloride	100	.274	1.375	0.59	.099	6.29	638	−142.0	103.6	3.21
	Trichloroethylene	100	.221	1.506	0.79	.073	7.20	477	−126.0	185.0	3.39
	Refrigerant 11	100	.199	1.556	0.61	.070	7.71	511	−168.0	74.8	5.19
−5	Calcium Chloride	25	.67	1.256	10.3	.29	2.85	513	−21.0	215.0	0.10
	Methanol Water	35	.89	0.962	9.9	.23	2.82	98	−22.0	176.0	0.51
	Propylene Glycol	50	.83	1.066	80.0	.23	2.72	60	−29.0	222.0	1.27
	Ethylene Glycol	45	.79	1.080	17.2	.25	2.82	103	−15.5	223.0	1.33
	Methylene Chloride	100	.273	1.394	0.68	.102	6.29	615	−142.0	103.6	3.26
	Trichloroethylene	100	.220	1.526	0.84	.076	7.16	479	−126.0	185.0	3.44
	Refrigerant 11	100	.198	1.580	0.70	.073	7.68	490	−168.0	74.8	5.27
−30	Calcium Chloride	30	.63	1.316	27.8	.28	2.90	110	−47.0	216.0	0.14
	Methanol Water	45	.80	0.962	18.0	.22	3.13	91	−45.0	171.0	0.66
	Propylene Glycol	60	.77	1.077	700.0	.21	2.90	55	−55.0	227.0	1.53
	Ethylene Glycol	55	.73	1.106	75.0	.212	2.98	93	−43.0	227.0	1.65
	Methylene Chloride	100	.272	1.426	0.80	.105	6.14	599	−142.0	103.6	3.33
	Trichloroethylene	100	.218	1.551	1.08	.078	7.11	432	−126.0	185.0	3.49
	Refrigerant 11	100	.196	1.611	0.88	.076	7.61	468	−168.0	74.8	5.37
−60	Methylene Chloride	100	.271	1.452	0.97	.110	5.98	572	−142.0	103.6	3.39
	Trichloroethylene	100	.216	1.580	1.38	.081	7.05	400	−126.0	185.0	3.56
	Refrigerant 11	100	.195	1.644	1.25	.079	7.41	416	−168.0	74.8	5.49
−90	Methylene Chloride	100	.270	1.494	1.22	.114	5.83	536	−142.0	103.6	3.49
	Trichloroethylene	100	.214	1.603	1.82	.084	7.01	367	−126.0	185.0	3.61
	Refrigerant 11	100	.200	1.663	2.15	.087	7.22	342	−168.0	74.8	5.55

h_b * — Coefficient of heat transfer between brine and surface (Btu/hr-sq. ft-F), at 7 fps velocity for .554 in. ID tubing.

Fluid Density (Lbs/Cu. Ft) = 62.4 X Specific Gravity.

Table 5.7. System Design Guide — Process Requirements.

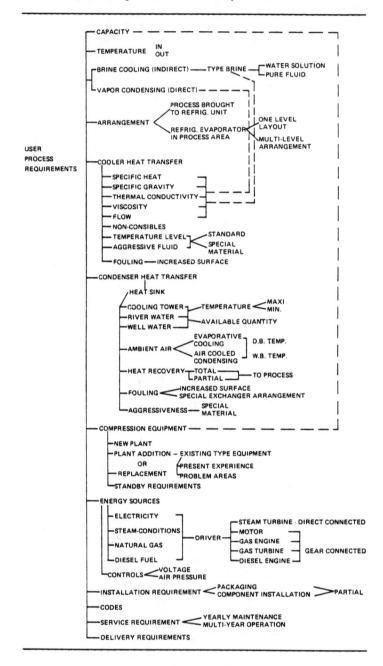

Table 5.8. System Design Guide — Designer Requirements.

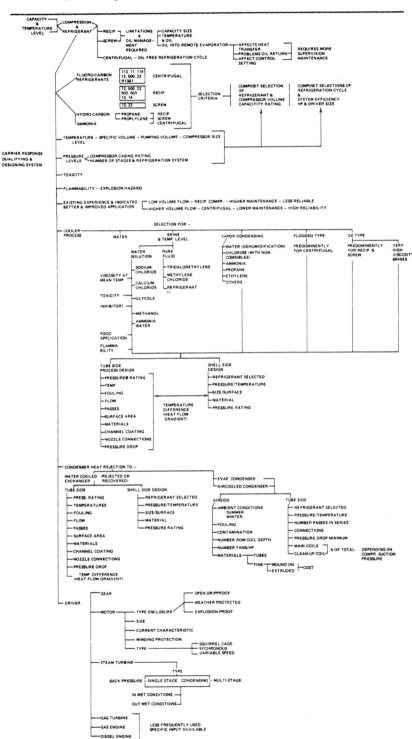

available and the decisions that have to be made by the refrigeration system designer.

For more detailed discussions of specific systems, design parameters and/or equipment, the following references should be consulted. The manufacturers of the various system components should also be consulted for specific equipment information.

REFERENCES

ASHRAE HANDBOOK & PRODUCT DIRECTORY — 1974 APPLICATIONS
  Food Refrigeration, Section III, Chapters 24–44
  Cryogenics, Section V, Chapter 50
  Refrigeration in the Chemical Industry Section, Section VI, Chapter 58
ASHRAE HANDBOOK & PRODUCT DIRECTORY — 1975 EQUIPMENT
  Refrigeration Equipment, Section II, Chapters 12–22
    Compressors, Chapter 12
    Steam-Jet Refrigeration Equipment, Chapter 13
    Absorption Air-Conditioning & Refrigeration Equipment, Chapter 14
    Condensers, Chapter 16
    Liquid Coolers, Chapter 17
    Liquid Chilling Systems, Chapter 18
    Component Balancing in Refrigeration Systems, Chapter 19
    Refrigerant Control Devices, Chapter 20
ASHRAE HANDBOOK & PRODUCT DIRECTORY — 1976 SYSTEMS
  Refrigeration Systems Practices, Section III, Chapters 24–32
    Engineered Refrigeration Systems, Chapter 24
    Liquid Overfeed Systems, Chapter 25
    System Practices for Halocarbon Refrigerants, Chapter 26
    System Practices for Ammonia, Chapter 27
    System Practices for Secondary Refrigerants, Chapter 28
    System Practices for Multi-Stage Applications, Chapter 29
ASHRAE HANDBOOK & PRODUCT DIRECTORY — 1977 FUNDAMENTALS
  Theory, Section I, Chapters 1, 2
    Thermodynamics & Refrigeration Cycles, Chapter 1
    Heat Transfer, Chapter 2
  Basic Materials, Section III, Chapters 15, 16, 17
    Refrigerations, Chapter 15
    Refrigerant Tables & Charts, Chapter 16
    Secondary Coolants (Brines), Chapter 17
HANDBOOK OF AIR CONDITIONING SYSTEM DESIGN
  Part 3 — Piping Design
  Part 4 — Refrigerants, Brines and Oil
  Part 7 — Refrigeration Equipment

CHAPTER 6

APPLICATION OF COOLING TOWERS
IN CHEMICAL PROCESSING

ALAN G. FURNISH
JOE BEN DICKEY, JR.
The Marley Cooling Tower Company
Mission, KS

GENERAL PRINCIPLES

The term "cooling tower" refers to an evaporative type cooling tower where water is circulated and distributed in direct contact with air. Therefore, most of the cooling is latent due to the evaporation of a portion of the water circulated. Cooling effect is increased by expanding the surface of water exposed to the air or by increasing the air velocity. The history of cooling tower development is in fact a documentation of the search for better ways to accomplish these two things.

Early efforts involved the use of lakes and then ponds constructed for this purpose. The exposed surface area was small and the air flow uncertain but some cooling was achieved. The spray pond was created by using a pipe distribution system feeding up-spray nozzles. This greatly extended the water surface exposed to air and achieved minor improvement in air flow giving much greater heat rejection per square foot of plan area.

Placing the spray system in a box open at the top and with fans that forced air in the bottom of the box resulted in the forced draft tower. The addition of fill to cause the water to splash and re-expose new surface to air flow gave us the basic components of modern cooling towers.

TYPES OF COOLING TOWERS (As Applied Today)

Mechanical Draft — Uses a fan to move ambient air through the tower. Mechanical draft towers are further described as induced or forced draft depending on whether the air is pulled or forced through the tower and Crossflow or Counterflow depending on the relative movement of the air to the down-flowing water. The induced draft wood mechanical draft crossflow cooling tower is by far the most prominent design in current use with an occasional counterflow as a special situation application.

Natural Draft — Depends upon a chimney or stack to induce air movement through the tower. No American application exists outside the power industry.

Coil Shed Towers — A combination structure of a cooling tower built over a shed structure which houses the tubular heat exchanging atmospheric coils or sections.

Spray Filled Tower — Solely dependent on spray nozzles for water break-up and therefore cooling surface. Very low thermal effectiveness compared to modern towers with fill. Occasionally used in very dirty or greasy services.

Spray Ponds, Atmospheric Deck Towers and Spray Towers (all depending on wind and natural convection for air movement) are seldom if ever used as performance is less than exact and land use excessive. They also drift badly in the wind.

DEFINITION OF TERMS

Heat Load — The quantity of heat to be dissipated by the cooling towers, usually expressed in BTU per hour. It is equal to the pounds of water circulated per hour times the cooling range.

Range — The number of degrees the water is cooled in the cooling tower (hot water minus cold water).

Approach (or approach to the wet-bulb) — The difference in temperature between the cold water leaving the tower and the wet-bulb temperature of ambient air.

Wet-Bulb Temperature — That temperature to which air can be cooled adiabatically to saturation by the addition of water vapor. More practically, wet-bulb temperature is the temperature indicated by a thermometer, the bulb of which is kept moist by a wick and over which air is circulated at the proper rate.

Dry-Bulb Temperature— Air temperature indicated by a dry-bulb thermometer.

Performance — The measure of a cooling tower's ability to cool water. Usually expressed in gallons per minute cooled from a specified hot water temperature to a specified cold water temperature with a specific wet-bulb temperature.

Drift — The droplets of circulating water carried out the fan cylinder by the exhaust air; usually expressed in percent of circulating water rate.

Blowdown — The intermittent or continuous wasting of a small amount of the circulating water. Its purpose is to limit the increase in concentration of solids in the circulating water due to evaporation. It is usually expressed in percent of water circulated.

Make-Up — The water required to replace the circulating water lost by evaporation, drift, blowdown and leakage. It is expressed in percent of water circulated.

Recirculation — A condition where a portion of the discharge air circulates back into the tower inlet.

Interference — The contamination of the tower inlet air with the discharge vapor of another cooling tower or heat source.

Cooling Tower Pumping Head — The total pressure required at the centerline of the tower distribution system inlet plus the difference in elevation of the inlet center-line and the top of the cold water basin curb. It does not include friction drop in the riser pipe.

SPECIFICATION

Most cooling towers are purchased on the basis of competitive bids. The Purchaser is responsible for issuing a specification with sufficient detail to clearly define the performance requirements, any siting limitations, information on water and air chemistry along with desired features, evaluation data and commercial considerations.

The following items should be considered for inclusion in purchase specifications.

Description of Service
GPM to be Cooled
Hot Water Temperature
Cold Water Temperature
Design Wet-Bulb Temperature
Pumphead Limitations (if any)
Water Analysis and Treatment
Site Wind Conditions (Velocity & Direction)
Design Wind Load — Structural
Seismic Design Requirements
Materials & Hardware Preference
Fan Hardware, Material and Number of Blades
Fan Stack Height
Motor Characteristics & Voltage
Evaluation Data
Guarantee Requirements
Plot Plan — Site Map Showing Tower Location, Material Storage & Access
Facilities Furnished by Purchaser
Installation Date
Commercial Conditions

EVALUATION FACTORS

With energy costs forecast to escalate more rapidly than the economy, cooling tower power requirements are an important factor in the economic purchase decision. Cooling towers for industrial processing (as opposed to power generation) tend toward widely varying design conditions and material requirements so the cost effect of lower horsepower in Figure 6.1 is not readily apparent in the "dollars per tower unit" columns.

Since many chemical plants are built on a short write-off period, usually 2 years, one might question the advisability of power evaluation. Capital costs are also rising and are tending to negate the effect of short-term operating power evaluation. The plant designer should particularly avoid singular concentration on fan horsepower alone. This usually leads to a high first cost selection not justified by a more accurate evaluation. A properly designed evaluation scheme can follow several paths from a straight factor comparison to a full present worth evaluation. However, the following operating costs should be considered in an appropriate manner:

Fan Horsepower — In many locations these may effectively apply only 3500 to

143

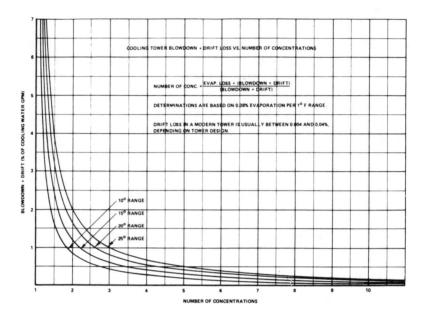

Figure 6.1

5000 hours a year. Fans may cycle on and off with varying heat loads and ambient air. Fans running at half speed only consume 1/7th of design horsepower but produce over 50% design air rate (cooling effect). Since the actual air entering the tower is less than design temperature about 98% of the time, two speed motors are a wise investment for minimizing operating costs. Where "coldest possible" water contributes to increased production, 8,760 hours per year evaluation time should be used. Power costs vary widely and of course local plant generated power is fairly common.

Pump Horsepower —

$$HPR_p = \frac{GPM \times Pump\ Head\ in\ Ft\ (approx.)}{3000}$$

Normally pumps are assumed to run 8,760 hrs per year even if fans are cycled. Standby pumps provide for maintenance outages. If the evaluation is to be meaningful, operating costs must be taken in context with pertinent installed costs and the total compared — one selection to another.

The following are significant installed cost evaluation factors:

Tower First Cost — Includes labor of installation.

Electrical & Controls — The cost of electric motors can be included in the tower materials price if so specified by Purchaser. Wiring and control costs should be evaluated for each cooling tower arrangement considered. Lower first cost will usually favor using fewer larger motors. Transformer costs can vary over a broad range on a small installation. This should be investigated through the Electrical Engineering Group or an Electrical Contractor.

Concrete Basin — The cold water basin and sump will cost between $10 and $15 per square foot of tower plan area assuming piling support not required. Piling can double this cost.

Piping — The cost of pipe, external to the tower will seldom influence the purchase decision. However, a detailed local evaluation may determine the best size, type and number of piping mains.

Finance — The cost of project financing can be viewed differently by different companies. Accounting assistance may be needed. In general, write-off periods are in the neighborhood of two years in the chemical industry. First cost, therefore, figures strongly in the annual "cost of owning the tower." Interest rates over the write-off period and during construction should be evaluated, also.

Figuring each factor and totaling will give a reasonably valid comparison between one selection and another. Of course, more sophisticated systems can be used and care should be taken that all factors are given proper weight.

PERFORMANCE CONDITIONS

Hot water temperature, cold water temperature and gallons per minute circulated are all interrelated (4). Geographic location will establish the design wet-bulb temperature. The process will determine the heat load. Range (hot water-cold water) can then vary with the water flow rate (GPM). To some extent, cold water temperature is established by the process served. However, the size and design of the heat exchangers can affect the cold water temperature required.

Cooling tower cost is particularly sensitive to cold water temperature. Small changes in cold water temperature make large differences in price thus careful consideration of heat exchanger design and size is necessary. Figure 6.2 will give typical conditions for guide lines. About 150°F should be considered a practical upper limit on hot water temperature since the structural strength of wood becomes affected as temperatures go higher. Wet-evaporative tower application above 160°F should not be commercially considered. The tower selection section will give some rating and economic guides to accomplish economic comparisons. Blow-down and make-up requirements relate to operating practice but provisions for them must be made in the original design. As water continues to circulate over the cooling tower, a portion of it evaporates rejecting the heat load to atmosphere. The water remaining will thus have a higher percentage of dissolved solids. With inadequate blowdown or makeup, this process can continue until scale forming tendencies appear, then some form of clean-up is required. The practical limit to the number

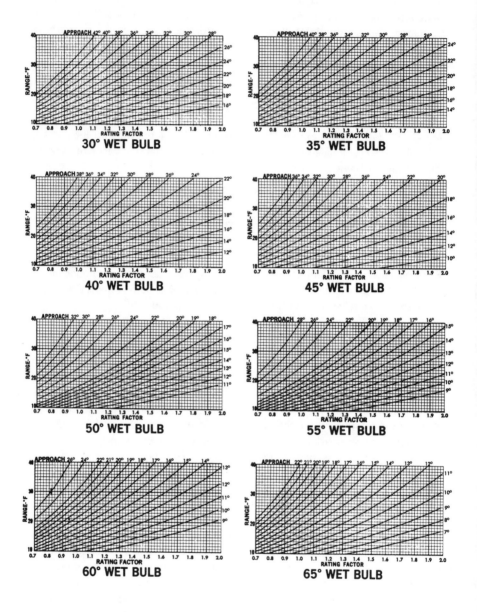

Figure 6.2 Crossflow mechanical draft water cooling towers typical rating factor curves.
(Interpolate between charts for intermediate wet-bulbs.)
(continued)

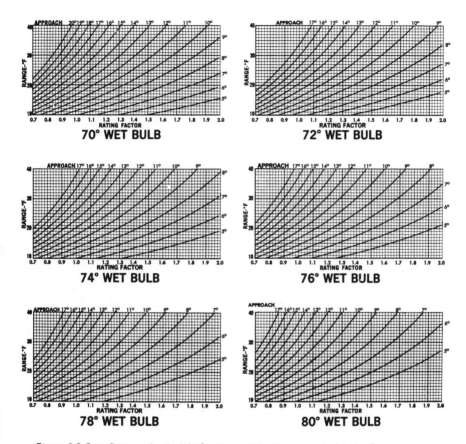

Figure 6.2 Crossflow mechanical draft water cooling towers typical rating factor curves.
(Interpolate between charts for intermediate wet-bulbs.)
(concluded)

of concentrations circulated depends on local water quality, on contamination due to heat exchanger leakage and on how dirty the plant atmosphere is (since the tower is a pretty good air washer). The chemical water treatment program should be referred to a competent water treatment specialist.

Once the allowable number of concentrations is determined then the blowdown rate required to maintain those concentrations can be calculated. Figure 6.1 will give the answer in terms of a percentage of the circulation rate to be drained to maintain the concentrations desired. Drift actually assists in this process of providing scale-free towers, but modern tower drift eliminators negate the need of calculation. In some areas, restrictions on availability of make-up water and effluent disposal make the use of low quality water for cooling purposes more desirable.

Today we see solutions approaching ph-11 in paper mill applications and phosphoric acid plants with circulating water ph-1.5 as examples of extremes considered. Dissolved solids concentrations can be quite high without harming the cooling tower materials or thermal performance to any extent as long as the ph is near the middle of the scale. Even sea water cools well and presents no insurmountable materials problems to the tower. Polluted tidal and brackish water are very corrosive and provide a changing chemistry that must be carefully evaluated when selecting cooling tower materials. Hydrocarbons that change surface tension and vapor pressure have drastic effect.

COOLING TOWER SELECTION AND BUDGET ESTIMATING

A system for rating mechanical draft cooling towers which has been in use for some time permits interpreting each application in terms of its relative degree of difficulty. The system is called the Tower Unit-Rating Factor method.

In this system the "Required Tower Units" are numerically equal to the Rating Factor times the GPM (gallons per minute of water circulation). TU = RF X GPM (see Figure 6.2).

Rating Factors define relative degree-of-difficulty. An example will indicate that a 1.1 Rating Factor is 10% more difficult duty than unity, requiring 10% more fill plan area compared to the area required at 1.0 Rating Factor.

This simple equation indicates that for duty conditions involving a Rating Factor of 1.0, the Required Tower Units would be numerically equal to the GPM. To further illustrate how this degree of difficulty function relates to actual selections, refer to the 70° Wet-Bulb Rating Factor Curve, and find that one Tower Unit will cool one GPM through any of the following combinations.

Range	9.8	13	16.7	21.2	26.8
Approach	8	10	12	14	16

The Rating Factor curves permit evaluating not only summer design conditions, but also cold weather operating conditions at normal or reduced water flow rates. Interpolation of data between tables is permitted, while extrapolation beyond the intended limits of the data is not recommended.

The Rating Factors provide a convenient means of interpreting relative degree of difficulty for the water cooling tower. All the most logical design conditions relative to industry are well within the limits of this data. Later, this discussion shows that cost is a function of "Required T.U." And, given "Available T.U." on a set selection, GPM and "R.F. Capability" are reciprocal.

The Rating Factor charts have established a method for determining the relative degree of cooling difficulty and have described this as a Rating Factor (1). Since the cost of the cooling towers will vary directly as the Rating Factor and since the cost of the tower will also vary directly as the GPM, it follows that the product of these two cost variables (the required Tower Units) is the truest cost function variable.

The Tower Unit will be recognized as a hypothetical square foot of heat transfer plan area used in much the same expression as a square foot of condenser surface. The plant designer or engineer, while optimizing, may accumulate a library of known costs, per-tower-unit and per-square-foot-of-condenser, as a means of projecting historical costs into the future, along with proper escalation factors. Since the ability of a tower to cope with a relatively more difficult or easier condition is quite similar to the same action in a surface condenser, it follows that in an evaluation problem they become reciprocal costs — more of one indicating less of the other.

If one is given a sufficient variety of historical tower costs with a statistically adequate distribution of rating factors, then a channel of costs-per-tower unit for equipment of equivalent specifications may be constructed. In Table 6.1 such a group of data is shown reflecting chemical and petrochemical tower costs for known projects current as of mid-1976 and projected for mid-1977 construction. This listing is rather complete for the period with deletions only where small size affected pertinence. Two special applications were left in to show the effect these requirements can have on costs. Most of the cost variation is due to differences in material specifications and not model to model costing differences.

These examples reflect the typical chemical plant performance condition of between 5°F and 10°F approach and between 20°F and 35°F range. However, holding certain of the variables constant, the reader may return to the Rating Factor charts and construct operating line curves at lower wet-bulbs, and/or different Rating Factor difficulty and/or similar manipulations including GPM and wet-bulb variations.

About half of the selections in Table 6.1 are low horsepower indicating power evaluation was part of the selection criteria. The estimator can only approximate this factor but it's adequate for evaluation at the estimating stage. This is fairly well proven by the fact that as BHP per T.U goes down, $ per T.U. will go up. This doesn't always happen in the Table 6.1 tabulation because differences in material specifications and labor markets can totally mask this trend. The tables and Rating Factors can be used as good budget figures. The finer points will come into line at bid time.

SITING CONSIDERATIONS

Chemical plants are rarely able to achieve a satisfactorily isolated location for a cooling tower installation. Ideally the cooling towers should not interfere with any other plant equipment nor should other plant installations affect the performance of the tower. Compromise being essential, let's look at the main factors to consider on both sides.

The cooling towers, being an air breathing device, ideally have a minimum of air restrictions, including a clean wind approach and departure path. Some compromise is necessary but pipe racks, for instance, are better neighbors than four story

Table 6.1. Actual Cooling Tower Selections for the Chemical, Petrochemical and Petroleum Industries. Gives Budget Estimate Pricing for 1977.

JOB NUMBER	DESIGN CONDITIONS							CALCULATED DATA					UNIT COSTS	
	10³ GPM	°F H.W.	°F C.W.	°F W.B.	°F RNG.	°F APP.	10⁸ BTU/Hr	FAN BHP	RATING FACTOR	NUMBER OF TOWER UNITS	BHP PER T.U.	10³ $ COST	$ PER GPM	$ PER T.U.
10	48.0	120	86	80	34	6	8.16	925	2.028	97,344	.0095	718	14.96	7.38
11	30.5	102	78	72	24	6	3.66	520	2.500	76,250	.0068	613	20.09	8.04
12	120.0	105	90	81.5	15	8.5	9.00	1040	.924	110,880	.0094	1144	9.53	10.32
13	60.0	110	85	76	25	9	7.50	800	1.408	84,480	.0095	634	10.57	7.50
14	40.0	105	87	75	18	12	3.6	600	.905	36,200	.0165	585	14.63	16.16*
15	28.0	110	87	80	23	7	3.22	429	1.465	41,020	.0104	410	14.64	10.00
16	40.0	110	85	78	25	7	5.00	752	1.636	65,440	.0115	476	11.90	7.27
17	18.0	98	83	78	15	5	1.35	242	1.582	28,476	.0085	266	14.78	9.34
18	50.0	120	88	79	32	9	8.00	788	1.471	73,550	.0107	550	11.00	7.48
19	41.0	114	90	83	24	7	4.92	336	1.370	56,170	.0060	543	13.24	9.67
20	72.5	115	85	75	30	10	10.88	1200	1.465	106,212	.0013	1260	17.38	11.86
21	24.6	116	90	81	26	9	3.20	285	1.237	30,430	.0094	346	14.07	11.37
22	11.0 Special	147	85	80	62	5	3.41	345	2.936	32,296	.0107	429	39.00	13.28*
23	33.0	115	90	84	25	6	4.75	344	1.525	57,950	.0059	551	14.50	9.51
24	20.0	110	85	80	25	5	2.50	450	1.990	39,800	.0113	320	16.00	8.04

Avg. Costs With Motors-Erected   .0093   15.75   9.82

Concrete Basin With Sump**   .97

Transf + Wire + Controls**   .85

   11.64

*NOTE: These two jobs represent rather special service and are not averaged in,

**"For Instance" Estimates

buildings. Downwind objects can cause serious recirculation even though the actual effect on inlet static pressure is small. Upwind objects, affecting the air flow pattern to the tower, can likewise cause recirculation. There are few rules here, only common sense and reliance on competent advice from your cooling tower Application Engineer are advised. Equipment which discharges heated or humidified air (including other cooling towers) should not be located upwind unless their effect is considered in the tower selection process.

Wind effects are mostly a matter of the plant geographic location modified somewhat by terrain and equipment layout. Some areas, such as the Texas Gulf Coast have very predictable prevailing winds. Other locations show almost no wind direction trends while most areas fall in-between. Several reference sources have statistically evaluated data from the National Weather Records Center in Asheville, North Carolina (5). A review of the frequency of occurrence of wind direction and velocity give good guidance on tower location and orientation.

As a general rule the tower should be oriented such that the prevailing wind is longitudinal (down the line of fan cylinders). Parallel lines of towers should be at least a tower length between louver faces to minimize interference. Fan stack height is particularly helpful here. 10' high stacks are much more effective than 6' and 18' is more effective than 10' at achieving a tower that gives good steady performance under varying wind conditions.

Drift is that portion of the circulating water carried out the fan stacks (2). Modern crossflow cooling towers have very low drift rates: .1% is standard, .05% easily obtainable and much lower drift rates are possible using reduced air rates and special eliminators. Zero drift is, of course, not possible thus consideration for effects of drift must be given in equipment layout, particularly for about 200 feet downwind.

Vapor Plumes

Cooling towers discharge heat in the form of water vapor to atmosphere. Condensation of this water vapor results in plumes, which merge together on multi-fan towers during cold weather. The tower may be located in a position which does not disturb downwind surface elements. This represents the normally least expensive plume management answer. A second approach is to elevate the discharge height of the fan cylinders. Approximately 100' above curb is the present maximum economical height, little if any ground level fogging will occur in the immediate area of the cooling towers. The added cost is about 35% of the tower price.

A third solution involves adding heat to the discharge effluent or, even more effective, reject some of the heat load with dry surface finned exchangers and mix the warm air from the coils with the wet section effluent to give a discharge mix below saturation (3). This resulting effluent will produce much shorter plumes. The part-visible plumes that may form will be quite diffused and do not tend to merge with adjacent cells. This wet-dry tower approach about doubles the cost of the cooling tower but can allow installations not otherwise acceptable. The ultimate solution is total dry cooling but that is another very expensive subject.

Schemes using gas burners in the fan stacks have been proposed and a few installations include this idea. The system will reduce or prevent fog but has several drawbacks like (a) combustion air adds additional moisture and (b) fire risk is great. Operating costs are impressive due to the quantities of gas consumed with the burners firing.

Sound

Sound is energy transmitted in the atmosphere in the form of sound pressure waves. The measurement of these pressure waves or sound pressure levels (SPL) is expressed in terms of decibels. Another characteristic of sound is frequency expressed as Hertz (cycles per second). The character of sound is analyzed in terms of these two functions.

To adequately evaluate sound the frequency range is arbitrarily divided into octave bands having center frequencies of 63, 125, 250, 500, 1000, 2000, 4000 and 8000 hertz. Standards of sound pressure levels at these octave bands and instrumentations to measure these SPLs give us the tools to identify and measure the level of sound sources individually and collectively.

Noise is objectionable sound. It is intangible and it is relative. A sound pressure level that may be acceptable to one person is an irritant to another. It doesn't take many complaints to result in an expensive remedy. It is necessary for the engineer to have an awareness of the potential problem and to analyze the proposed plant site to determine the acceptable sound pressure level both for the plant neighborhood and those working in the plant, as well as OSHA.

Cooling tower location and orientation should be considered first. Placing the towers away from possible complainants is usually the least expensive approach. Placing the tower's endwall (non-louvered face) toward the most sensitive area will also help as the following data shows.

Typical Sound Pressure Levels — Mechanical Crossflow

(db, re.0.0002 microbar in free and unobstructed environment)

	dBA	dBC	63	125	250	500	1000	2000	4000	8000
					Band Hz					
50' Louver Face	72	83	77	76	75	68	65	63	62	63
50' Cased Face	59	73	66	63	62	57	53	46	35	31

MATERIALS OF CONSTRUCTION

Structure — Cooling Tower structures are constructed of wood, steel or concrete depending on tower design and application. For process, chemical and petrochemical application the treated wood structure remains unchallenged. The treated wood most used today is West Coast Douglas Fir, largely for economic reasons. Fir costs tend to vary with the housing market but continue to demonstrate the ability to remain market-competitive. California redwood, while still an excellent structural wood, has posed intermittent procurement problems. Treated fir has, in 30 years of use, proven its ability to duplicate the longevity of treated redwood and, will far outlast untreated redwood in the average application.

Treated fir plywood is also used extensively in cooling tower construction. Exterior grades are, of course, required. Main applications are for fan deck and hot water basin flooring.

The treated wood structure offers many years of trouble free service with conditions of varying water chemistry. Since water chemistry is increasingly a problem and also subject to upsets, most chemical plant cooling towers start with premium hardware such as stainless steel.

Modern towers utilize structural connectors of noncorrosive molded fiberglass or structural plastics. This keeps to a minimum the amount of ferrous material adjacent to wood structure and reduces the brown cubical rot potential.

Fill — Cooling tower fill materials vary. Treated wood continues to dominate with a combination of properties such as chemical resistance, structural strength and cost. Polyvinyl chloride is extensively used and offers excellent chemical resistance. Its fire retardant properties are a plus. The structural strength of PVC is low and the cost presses users to take all advantage possible of the available strength. Polypropylene offers both excellent chemical resistance and a cost relation which challenges wood. Polypropylene is a good substitute for wood in cooling tower fill for chemical plant use.

Mechanical Equipment — This is the heart of the cooling tower. It must be ruggedly designed for continuous operation under corrosive conditions. It must, in other words, be designed and developed specifically for cooling towers use. Some manufacturers of cooling towers design, test and manufacture their own mechanical equipment, providing a single responsibility for all tower components (except motors).

Fans — Propeller type fans are used exclusively on process cooling towers because they can deliver large volumes of air against low static pressures with very high efficiency — 80% of more in larger sizes. Solid cast aluminum alloy and fiberglass fan blades dominate both the original equipment market and the replacement market. Special consideration can be given fan construction where particularly corrosive conditions are encountered.

Fan blades of compression molded fiberglass are available in all common sizes (up to 30' diameter) and most likely will be the sole material used as larger sizes are developed. Fiberglass is standard for fan hub covers. Attaching hardware is galvanized steel but premium materials of most any alloy can be supplied. Fan hubs are either cast iron, aluminum plate, or steel weldment design depending on fan diameter. Coal tar epoxy coatings, properly applied, now give excellent corrosion protection.

Speed Reducers — Requirements are stamina and long life. Speed reducers must be ruggedly constructed for continuous service in the severe environment in which they operate. Several types of gearing are used. Spiral bevel, helical and worm gears are most common. Depending on size and reduction ratio required, the speed reducer may use a single type or a combination of any two types. Generally, two-stage units are desirable for 20′ and larger fans.

The service life of a gear box is directly related to the surface durability of the gears. The American Gear Manufacturers Association has established service factors that apply to cooling tower installations. This factor is the ratio of calculated basic horsepower to the applied horsepower. It varies with the type of prime mover and the type of duty — intermittent or continuous. With electric motor drive, a service factor of 2.0 on spiral bevel gears and 1.5 for helical on continuous duty are widely accepted.

Bearings are normally selected for a calculated service life compatible with the type of service. For continuous duty, the established standards specify a B-10 life of 100,000 hours. B-10 life is defined as the life expectancy in hours during which 90% or more of a given group of bearings under a specific loading condition will still be in service.

Lubrication is highly important to long trouble free service life. The system should be a simple design and capable of at least limited periods of reverse operation (for cold weather de-icing of louvers). A slinger or splash system adequately meets these requirements and is not subject to wear. Static oil level should cover the gears for corrosion protection during shut down. Oil level indicators outside fan cylinder and conveniently located drain and sampling lines simplify good preventive maintenance. Of course, lubricants and procedures should carefully follow manufacturer's recommendations.

Driveshaft — A driveshaft of simple, rugged, full floating design is required to transmit power from the motor (located outside the fan cylinder) to the gearreducer. Simple implies non-lubricated for minimum maintenance; rugged because it operates in a corrosive environment and full floating design for its proven service record. Flexible coupling elements of molded neoprene bonded bushings give excellent service, are practically impervious to corrosion and are easily replaceable. Couplings of galvanized castings or stainless steel with stainless tube and flange assemblies are commonly used and recommended for this critical item.

Dynamic balancing is a must for driveshaft assemblies to reduce vibration forces and keep bearing loads to a minimum.

Enclosure — Cooling tower casing and louvers have utilized 3/8″ corrugated ACB (asbestos cement board) as standard construction for over 20 years. Lapped horizontal joints and lapped and mastic-sealed vertical joints give a good watertight enclosure on endwalls. Structurally supported on 4' centers, the louvers will carry a heavy ice load even if de-icing cycles are too infrequent.

In sea water service, unprotected ACB can accumulate internal salt deposit with attendant swelling and spalling. Inside coating using a mastic or PVC material can prevent this. Low ph can shorten the service life of ACB such that wood fiberglass casing and louvers may be the only answer at some extra expense. Service life of fiberglass casing should equal ACB with proper care in installation.

DESIGN FEATURES

The following design features contribute to the utility and lower maintenance costs associated with a wood mechanical draft cooling tower.

Freezing weather operation must be expected to some degree in most areas of the United States. Ice control is very much a matter of operating procedure but certain design features gives a greater margin for error. The crossflow tower was an improvement over counterflow for two reasons: (a) A much broader flow range permitted overpumping a portion of a tower to prevent or remove ice formation, (b) splashout and the attendant ice on the louver face die not completely block air flow letting cooling proceed over much greater time cycles. Since 1959, the sloping louver face design of crossflow towers has added a new dimension to ice control. Large wide spaced louvers are used along with a fill positioning which carries water to the inside portion of these louvers. This washing action keeps ice from building up and blocking air flow, letting the tower continue to cool water. By first shutting off fans and then reversing fans, heated air and water can be moved out on the louvers, thawing ice. Full speed fan reversal is recommended for de-icing. On any cooling towers, it is important to keep maximum water flow and heat load during freezing weather (staging fans off, one cell at a time to maintain a warm water temperature).

Dirt and debris can collect in and block distribution systems. Algae can grow in fill and particularly on eliminators where air flow is blocked. Occasionally fan blades can collect deposits. Most system dirt collects where water moves the slowest — in the cold water basin. Periodic inspection and clean-up are necessary on any tower. The time interval will vary — mostly with the amount of airborne dirt since the tower is a good air washer. Algae can usually be controlled with chlorine shock treatment. Residuals should be less than 1 ppm or tower lumber will suffer.

Piping within modern cooling towers can be set up in several ways and made of different materials. Cast iron pipe is most commonly used in small sizes, up to 24" diameter. Fiberglass pipe, especially the RPM (reinforced plastic mortar) construction is fast taking over in the larger sizes. For individual cell inlets, cast iron is preferred since it handles economically and is very durable. On multicell towers a single or double end inlet system of RPM or steel are usually used because the piping costs are significantly less expensive. Strong, durable, low pressure drop flow

Figure 6.3 Ten models of this factory-assembled cooling tower offer a wide range of cooling capacity. Modular construction permits easy expansion. (Courtesy: Marley Cooling Tower Co., Mission, KS.)

Figure 6.4 Double-Flow towers in single-cell capacities from 600 to 900 nominal tons, this tower provides more tons of cooling per square foot of plan area with less energy consumption than any other factory-assembled cooling tower of comparable capacity. (Courtesy: Marley Cooling Tower Co., Mission, KS.)

Figure 6.5 Hyperbolic Natural Draft Towers, like the round mechanical draft towers, are suitable only for the very largest heat load applications. Hyperbolics may be specified, 1) where geography dictates maximum height release for moist air plume; 2) for confined, restricted area sites; 3) where extremely high power costs make natural draft most applicable. Attendant advantages of Hyperbolic towers are simplicity of operation, since no mechanical equipment is used, and total avoidance of heated air recirculation, as plume dispersion is of considerable height. Available in cross-flow and counter-flow design.

control valves should be used for flow balancing over the tower and provide tight shutoff for occasional cleaning and maintenance. Coil towers, or cooling towers built on top of a coil shed, are still occasionally used but economics strongly favor using a standard cooling tower in conjunction with a shell and tube heat exchanger. The coil shed is used to structurally support the weight and wind loads of the cooling tower as well as house the atmospheric coil sections and water redistribution system. In general using a coil shed design will add 30%–50% to the system cost.

REFERENCES

1. Dickey, J. B., Jr. and R. E. Cates, "Managing Waste Heat with the Water Cooling Tower," 2nd Edition, The Marley Company, 1973.
2. Holmberg, J. D. and O. L. Kinney, "Drift Technology for Cooling Towers," The Marley Company, 1973.
3. Hansen, E. P. and R. E. Cates, "The Parallel Path Wet-Dry Cooling Tower," The Marley Company, 1972.
4. Maze, R. W., "Practical Tips on Cooling Tower Sizing," The Marley Company, 1967.
5. Army, Navy and Air Force Manual, Engineering Weather Data, U.S. Government Printing Office—651328, 1963.

CHAPTER 7

INDUSTRIAL STEAM TRAPPING

JACK F. CURRAN
Yarway Corp.
Blue Bell, PA

INTRODUCTION

We have merely to look around us, these days, to be reminded of the significance of industrial steam traps. As long as steam cost only 50¢ or 55¢ per M lb steam traps, once installed, could be promptly forgotten. As the cost of steam rose to its present levels, however, it has become imperative to install the right traps properly, and to make them a part of the plant's regular maintenance schedules.

A large percentage of problems involving steam traps arise from a misunderstanding of the correct function, operation, or application of the steam traps that are being used. Carefully selected and properly installed, steam traps provide an important contribution to the efficient operation of a steam system and in saving energy.

A basic steam system starts with a boiler into which treated water is pumped. Heat and pressure are then applied to convert hot water into steam. Steam is then transmitted through a steam main, and distribution and branch lines, to heat steam users such as steam tracing lines, heating units, and process equipment.

Steam traps, which should be installed at the natural drainage points of all steam users, discharge to a collecting system which returns the condensate to a receiver or flash tank. Condensate is then mixed with treated make-up water and pumped back into the boiler to generate more steam. Due to the substantial amounts of steam that can be conserved, more and more steam systems now provide for collecting and returning condensate to the boiler for reuse.

MAJOR FUNCTIONS OF STEAM TRAPS

The ideal steam trap should perform three major functions:

It should discharge condensate quickly, the moment it forms, but hold back the steam for useful work.

It should eliminate all air and gas from the system quickly, particularly on startup after equipment has been shut down for a period of time.

It should accomplish both of these tasks even in the face of changing pressure or condensate load conditions in the line ahead of the trap.

Prompt drainage of condensate is essential, since if condensate is not removed, the steam spaces will soon become waterlogged and steam will not be able to reach the heat transfer surfaces.

158

The second important function of a steam trap is the removal of air and non-condensable gases that may be carried long the lines and into the heating equipment.

Air may enter the system either through the boiler make-up water, or it may be sucked back into the lines and equipment on shutdown, due to condensation of the steam. Other non-condensable gases such as carbon dioxide (CO_2) and carbon monoxide (CO) can be formed in varying amounts of chemical action in the boiler. This depends on the kind and amount of boiler water treatment that is employed.

In contrast with steam, air and gas are non-condensable. Therefore, they are carried along the lines with the steam into the process equipment or heating coils. Air and gas tend to form in layers around the heating surfaces and, together with the condensate, act as insulating blankets.

In addition to this insulating effect, some of the air and gas mixes with the steam, lowering the temperature of the resulting mixture. This slows down the rate of heat transfer and further reduces heating efficiency.

From this it is evident that not only must condensate be removed as quickly as possible from the heating equipment, but that air and gas must be removed as well to maintain temperatures and thermal efficiency.

FLASH STEAM

Flash steam discharged from steam traps is often mistaken for live steam and the traps are blamed for "blowing" or wasting steam, even though the traps in question may be operating in a completely correct and efficient manner.

An understanding of the thermodynamic phenomenon involved will help provide a better understanding of correct steam trap operation.

When hot condensate is discharged to the atmosphere or from a higher pressure to a lower pressure in a condensate return line, it contains more heat than it can hold at the lower pressure.

For example, take a steam trap operating at 100 psig and discharging to atmosphere. Steam tables will show that the heat of the liquid for this pressure and saturation temperature is 309 BTU. This is also the amount of heat in one pound of water just before it turns to steam at 100 psig and 338°F. So, when this pound of steam first turns to condensate, it will contain 309 BTU at this temperature and pressure.

Steam tables also show that the heat of the liquid at 0 psig or atmospheric pressure, is only 180 BTU per pound, and that the steam temperature for this pressure is 212°F.

Consequently, when the pound of condensate containing 309 BTU at 100 psig is suddenly discharged to atmospheric pressure, or 0 psig, it can only retain 180 BTU of heat. Therefore, the excess heat (difference between 309 BTU and 180 BTU) practically explodes a portion of the condensate which immediately boils off as flash vapor, or flash steam at 0 psig.

This flash steam is often mistaken for live steam loss produced by malfunction-ing steam traps. But actually it is the perfectly normal and unavoidable result of the discharge of hot condensate from a higher pressure to a lower pressure. A photo of a trap properly discharging hot condensate and flash steam is shown in Figure 7.1.

Figure 7.1 Trap properly discharging hot condensate.

Flash steam will also be formed in the same way if the traps discharge into a low pressure return system. In such a case, the volume of flash steam will vary with the pressure in the return system. The higher the return line pressure, the smaller will be the volume of the flash steam formed.

Note that unless the return lines and condensate receiver are large enough to handle the flash steam in addition to the condensate, the return line and receiver tank pressure may rise above the intended figure. This can adversely affect trap operation, overload the receiver tank, and may cause a loss of live steam and excess flash steam out of the receiver vent.

Flash steam plays a vital part in the operation of the thermodynamic class of steam traps. They open and close through the proper utilization of the pressure variations resulting from the flash steam formed in the control chambers and orifices of the traps.

The nearer to steam temperature and the more quickly a trap discharges the condensate that comes to it, the greater will be the percentage of flash steam that will be formed.

By discharging condensate quickly, as soon as it forms, a steam trap keeps equipment operating at a higher temperature and therefore at a greater efficiency. Consequently, the trap that will discharge the condensate the quickest will also produce the greatest percentage and volume of flash steam but at the same time, it should prove to be the most efficient.

It is quite true that there is useable heat in the flash steam. But it is heat that cannot be used at the operating pressure in the process apparatus if the desire is to get rid of the condensate as quickly as it forms in order to promote efficient heat transfer.

It is, of course, possible to make use of the heat in the flash steam at some pressure lower than the original. For example, if process steam is being used at 100 psig, the steam traps draining this system could be discharged to a 15 psig process line, or to a low pressure heating system. The flash steam could also be used to heat the boiler feedwater or to supply heat to the hot water generator. In fact it could be used anywhere in the system where low pressure steam is needed.

When traps are discharging to the atmosphere it is usually quite easy to determine whether the vapor arising from the discharged condensate is live steam or merely flash steam.

If the discharge issues from the traps as a strong jet or blast and is colorless close to the point where it issues from the discharge pipe, some live steam is probably being lost. An all white discharge with no clear jet usually means that it is only flash steam mixed with the condensate. Similarly, vapor rising from the trap discharge in a lazy drift or cloud represents flash steam rather than live.

An understanding of the difference between flash and live steam is helpful in trouble-shooting traps. This subject will be discussed later.

Flash steam can be minimized in outdoor installations where traps discharge to the atmosphere by connecting the discharge lines to a manifold or outdoor flash tank. Radiation of heat from the manifold will condense the flash, reducing the volume of the vapor which arises from the discharged condensate.

Of course, it is possible to have a trap that will hold the condensate back until it cools below the boiling or flashing point. For most installations this would not be advisable, since condensate would be backing up in the apparatus, blanketing the heat transfer surfaces and lowering the efficiency of the operation.

THE BASIC TYPES OF STEAM TRAPS

Mechanical — These traps operate by responding to the difference in density between the water and steam that come to the trap. They open to water or condensate, but close on steam.

Thermostatic — They are actuated by the temperature of the fluid reaching the trap. They open on cool condensate, but close off when condensate temperature approaches the temperature of steam.

Thermodynamic — This type of trap uses the difference in the amount of energy available in cool condensate, hot condensate, or steam to open or close the valve mechanism of the trap.

Inverted Bucket, open bucket and float are the three basic types of mechanical traps. The inverted bucket design supersedes the open bucket type and is the most commonly used mechanical trap at present. It is somewhat similar to the Open Bucket trap, but the bucket (A) is inverted and is open at the bottom. A valve linkage mechanism (B), attached to the top of the bucket, permits the discharge valve (C) to open and close as the bucket falls and rises. Bucket traps should always be primed to prevent loss of steam on initial startup (see Figure 7.2).

Thermostatic traps are actuated by the temperature of the fluid reaching the trap. They open on cool condensate, but close when condensate temperature approaches the temperature of steam. Most thermostatic traps can be mounted to discharge vertically downward, so the trap is self-draining. This can be important where protection against freezing is desired. The operating element consists of a corrugated bellows (A) mounted within a housing, usually of cast iron. A valve (B), mounted at the bottom of the bellows, closes the orifice (C) when the bellows expands (see Figure 7.2).

Float and Thermostatic traps combine the features of the mechanical type with a thermostatic element similar to that used in thermostatic traps. It is intended for low pressure applications where large volumes of air must be eliminated in addition to the condensate. Float controls condensate discharge, thermostatic element controls air elimination.

Dual Range Thermostatic trap uses both thermostatic and thermodynamic principles to provide wider range and faster response than conventional thermostatic traps. Series 151 trap has two valves. The pilot valve, which functions thermostatically, responds fast to changes in temperature, and can handle light rates of condensate. The main valve, which functions thermodynamically, stays closed until the load exceeds the capacity of the pilot valve. Then both valves discharge at full rating.

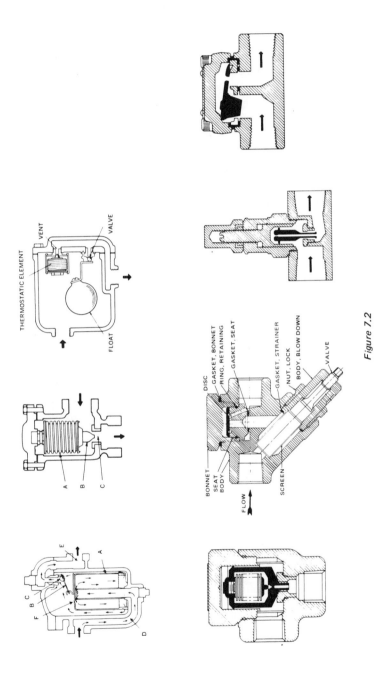

Figure 7.2

Thermodynamic traps utilize the flash steam developed during the discharge of hot condensate to control the opening and closing of the trap. One of the most widely used types of thermodynamic traps is the disc type. It is essentially a time cycle device ... in other words, under normal operating conditions, each time the disc closes on steam at a given pressure and temperature, it will stay closed for approximately the same length of time. The trap reopens when the steam, in the bonnet chamber above the disc, condenses sufficiently to permit inlet pressure to push the disc open. In its smaller sizes this trap is widely used on light load applications such as steam line drips and tracers. Larger sizes are used on heavier condensate loads. The piston valve type of trap consists essentially of a piston type valve operating within a control cylinder. The lower end of the valve has a tapered seating which opens and closes the orifice. The lever valve type of trap (shown) operates on the same basic principle as the Piston type, but with a lever action, rather than a piston. It is designed for extra heavy high temperature condensate loads (see Figure 7.2).

All thermodynamic traps can be used on systems where non-freeze installation is essential.

CONDENSATE RETURN SYSTEMS

The large volume of flash steam that must be carried off with the condensate is an important factor often overlooked in sizing return lines.

Back pressure can build up in the return lines, perhaps even to the point where it restricts traps discharge rates or prevents some of the steam traps from closing off properly. If this happens, it will only aggravate the problem by allowing live steam to enter and further overload the returns.

Should this occur, the receiver tank may be overloaded, the condensate pump may become steam bound, and live steam may be lost out the receiver vent.

To avoid this problem, provision should be made when designing the piping layout to allow sufficient capacity for the additional volume required by the flash steam as well as the condensate.

The discharge line from each trap should be at least as large as the pipe size of the trap it is serving. Where two or more traps are discharging into a common return, the pipe size of the return should have an internal area approximately equal to the sum of the areas of the individual lines feeding into it.

The receiver tank should also be large enough to accommodate both the condensate and the flash steam. The flash tank should be vented to the atmosphere in order to avoid raising the return line pressure above the desired limit and perhaps producing excessive back pressure on the steam traps and return system.

BACK PRESSURE PROBLEMS

A frequent cause of high pressure difficulties in return lines is the installation of additional steam equipment in a plant, without the corresponding adjustments to return line capacity.

The initial reaction to such a situation may be to blame the steam traps. But in fact the overloaded return lines and receiver are actually the culprits. In such a case, the correct remedy is to analyze the return piping system and enlarge and update it.

It is, of course, advisable to check over the steam traps at the same time to ensure that they are all operating properly. Even a few traps sticking open and passing live steam into the returns can greatly magnify the problem. Also check for open or leaking bypass valves as a trouble source.

WATCH FOR THESE KEY FACTORS IN PIPING HOOKUPS

A correct piping hookup is just as important as the correct choice of steam trap. Both factors must be given careful consideration for best heat transfer and most efficient steam trapping. Key considerations are illustrated in the diagram above.

1. Branch steam supply lines should always be hooked up to the top of the steam main. This prevents condensate from being picked up from the main and carried over to the heating equipment.

   Slope horizontal run of supply line slightly toward main so that any condensate formed there will drain back into the main rather than into the equipment. Make sure supply line is large enough to allow full flow to equipment without pressure drop.
2. Traps draining supply lines should be located just before the line enters the equipment, if the line is more than a few feet. The line should enter equipment as near the top as possible.
3. Condensate discharge lines should be attached to the equipment at as low a point as possible to prevent pockets of condensate from accumulating and blanketing heating surfaces. Size of the line should be at least equal to pipe size of the trap. Slope the discharge line toward trap to avoid water hammer.
4. Valves and unions should be placed on either side of the trap to simplify trap servicing. Uniform spacing between unions for all traps on the same type of service allows easy removal, and insertion of a spare assembly while servicing is done in the shop. All valves should be gate or ball type to avoid pressure drop.
5. Overhead discharge, where a trap must discharge against a lift to an overhead line, requires a check valve at the bottom of the lift to prevent back flow when equipment is shut down. An external check valve is not necessary for some traps that incorporate their own check valve feature. Attach the discharge line so condensate enters at top of the return main. Pressure due to lift plus pressure in the return main should not exceed the back pressure limitation recommended by the trap manufacturer.
6. Test tees installed just ahead of the shutoff valve to the return line provide fast, easy trap testing when discharge is into a return system. A T-fitting with a test valve is installed, so by closing off the shutoff valve and opening the test valve, the discharge from the trap can be observed.
7. Strainers should be installed just ahead of all steam control devices, including steam traps. Unless the trap is equipped with an integral strainer, a strainer with a blowdown valve should be installed just ahead of it. The relatively small cost of the strainer will be repaid many times over in protection from dirt.

   Some thermodynamic and thermostatic traps can be repaired in-line. Upstream unions which are a major source of steam loss when not tight, can be eliminated.

TRAP APPLICATIONS

Industrial steam traps fall into three application categories — non-process, process, and environmental heating.

Non-process applications are characterized by high condensate loads; relatively constant pressures, temperatures and condensate loads; and small amounts of air to hand. Typical non-process applications include steam main drips, steam tracing, and small unit heaters.

Process applications, on the other hand, are characterized by heavy condensate loads; fluctuations in pressure, temperature, and loads; and usually have large amounts of air to hand, particularly on start-up. Typical examples are shell and tube heat exchangers, air blast heaters, water heaters and vaporizers.

Environmental heating applications include air heaters, unit heaters, and finned tube wall heaters for space heating.

Trap Selection

Here is a list of the criteria used in selecting the type of steam trap for a certain application:

Ability to remove air from the system

Ability to match the operating pressure conditions of the system

Proper discharge capacity

Response to variations in temperature, pressure, and condensate load

Selection on the basis of physical ruggedness, ease of installation, or resistance to freezing. Out of doors, a trap that holds water in its body could freeze and jeopardize the entire piping system

It is also important to consider operating pressure in selecting the trap. Traps have often been misapplied through sizing based on pipe size in the system, or pipe size of the trap, rather than the relationship of the trap to the load involved.

Trap Sizing

Trap sizing and safety load factors are important considerations. Too large a trap may cause sluggish operation and waste steam. Too small a trap will back up condensate and impair the efficiency of the heating operation.

Safety Load Factor

The safety load factor, or load multiplying factor, is one of the essential considerations in calculating the proper trap size. It is simply a multiplying factor to be applied to the expected condensate load to:

Allow for larger amounts of condensate formed by cold metal on startup as well as condensate load fluctuations during operation.

Provide additional trap capacity needed for air venting.

166

Allow for below normal operating pressure and resultant lower trap capacity.

Allow for possible inaccuracies in estimating condensate load.

Take into consideration the process equipment operating characteristics as compared to trap characteristics e.g.—continuous or intermittent condensate flow vs type of trap discharge operation; steam binding; or modulating steam controls.

Range of Safety Load Factors

Since some variations in condensate load must be provided for in every case (due to startup, pressure or temperature fluctuations, and/or process load changes) a safety load factor of at least 2 should always be used.

In cases where wide variations of condensate load or operating steam pressures are encountered, as with modulating control valves, a safety load factor of 4 would be in order.

For certain applications, and for some types of traps, where especially large volumes of air must be handled, an even larger factor should be used, perhaps as high as 8 or 10. Typical examples of such applications would be cylinder dryers or large autoclaves.

When involved in an application where the proper safety load factor is in doubt, consult with the manufacturer of the traps that will be used.

Some trap manufacturers provide trap selection tables which already take into consideration the applicable safety load factor in listing the recommended trap sizes. In such cases, no further consideration of the safety factor is required.

Determining the Trap Size

1. Find out the actual pressure at the trap inlet (maximum and minimum). This is frequently considerably less than the supply line pressure, for pressure drop takes place not only in the process equipment, but sometimes in the feeder lines as well.
2. Determine the maximum and minimum pressure at the trap outlet. Make sure that the ratio of outlet pressure to pressure at the trap inlet does not exceed the limitations recommended by the manufacturer of the trap being used.
3. Obtain the figure for the maximum condensate load arriving at the trap. If this is not known, some calculations will be necessary. Determine the operating conditions of the heat transfer apparatus, i.e. the production rate, the flow rate, temperature rise of the process, the holding or final temperature of the process material, or any other factors from which steam consumption may be calculated.

Condensate Temperature

Trap capacity is directly affected by the temperature of the condensate flowing through the trap orifice. The hotter the condensate, the lower will be the capacity of a given trap orifice. Capacity of an orifice on condensate at steam temperature will only be one-quarter to one-fifth the capacity of that same orifice on water or condensate at 200°F (just below the atmospheric boiling temperature).

Since condensate temperature will be about 30°F below steam temperature in most applications, the trap capacity will generally be about half as much as it would be for relatively cool (200°F) water or condensate.

For example, a hot water generator where the incoming water is relatively cool might form condensate at approximately 30°F below steam temperature. On the other hand, a trap on a steam main drip, or on equipment such as a cylinder dryer might be required to handle condensate very close to steam temperature.

Manufacturers' trap selector tables incorporate a factor that takes into consideration the probable condensate temperatures as one of the variables on which the recommendations for trap sizing are based.

Trap Location

Traps should be located below the equipment to provide gravity drainage toward the trap inlet. A horizontal drain line should be pitched toward the trap to provide positive hydraulic head at trap and to prevent water hammer. In order to facilitate maintenance and servicing, the trap should be placed in an accessible location.

If condensate must be discharged over the side of a tank or vat, a lift fitting or water seal should be used to prevent steam binding of a trap. This can be either a U-shaped loop, or if the coil in the tank is large relative to the size of the discharge line to the trap, a small pipe within the steam coil and reaching down to the bottom of the loop may be used.

Clean Out Piping

Thoroughly blow out piping to clear out scale and debris before attaching the strainer and trap. If this is not feasible blow out the strainer screen under full steam pressure after installation.

Overhead Discharge

When a trap is connected to an overhead return, pressure at the outlet of the trap must be sufficient to overcome the lift (calculated at 1/2 lb per foot) plus the pressure in the return line. Adding pressure due to lift to the pressure in the return line gives the total back pressure to be overcome. Be careful to make sure that this total back pressure is within the allowable limit recommended by the manufacturer of the trap being used.

Multiple Coils or Rolls

Where air is being heated as it passes through multiple banks of coils, or material is being dried by passing it over rotating drums, the first coil or drum will form the heaviest condensate load. Therefore, traps on the first and possibly the second coil or roll should be larger than those draining the others.

Unit Trapping

Trap each coil or roll separately. If more than one coil or unit is hooked up to a single trap, short circuiting is very likely to occur due to differences in pressure drops. The coil or unit with the least pressure drop will blanket or short circuit the others and cause uneven and inefficient heating. Group trapping can also result in incomplete air removal and water logging of some of the units in a group.

Piping to and from the traps should be at least as large as the pipe size of each individual trap.

Common Return

Where several traps discharge close together into a common return, install a swing check valve between each trap and the return. Otherwise reverse flow, when one or more units are shut down, or the blocking off of some units may occur as traps discharge. The return should be of ample size.

By-Passes

Due to risk of steam loss when left open unnecessarily, by-passes are seldom used in modern plants. Where shutdown for even a few minutes cannot be tolerated on a critical installation, the use of two traps in parallel may be substituted for a by-pass. If a by-pass is necessary make sure it is not left open when not needed.

If used with a bucket trap, arrange the by-pass piping so it will be at a higher level than the trap to avoid losing the prime in the trap.

Bucket Traps

To prevent loss of steam on initial start-up, bucket traps should always be primed. To do this, close outlet valve at trap and open inlet valve slowly to allow condensate to fill trap body. If insufficient condensate comes to the trap pour in water through the test outlet in the cover of the trap.

Vacuum Breaker

Where specified by the equipment manufacturer or where condensate might be sucked back into equipment on shutdown, install a vacuum breaker at a high point on the apparatus.

Auxiliary Air Vent

Where large volumes of air must be eliminated on start-up, install an auxiliary air vent at a point on the apparatus opposite the steam inlet.

Freeze Protection

Most thermodynamic and thermostatic traps can be mounted to discharge vertically downward to avoid freezing. Keep discharge lines short and pitch downward to permit gravity drainage. To keep the strainer in a horizontal line from freezing, turn it on its side.

Insulate discharge lines from lightly loaded traps and in all cases if the discharge line is long or runs vertically to an overhead return. Provide for drainage of risers on shutdown. Always insulate bucket traps and provide for drainage of traps if lines are subject to shutdown.

Strainers used in out-of-door applications should be positioned to prevent freezing. This can be done by mounting the strainer on its side in horizontal lines and vertically in vertical lines.

Maintenance

Trap maintenance involves a few simple steps such as cleaning and replacing parts. All trap manufacturers supply maintenance and repair instructions for their traps. When carefully followed, these instructions assure the efficient operation of the traps. Maintenance inspection can reveal, however, that a trap may not be functioning properly. Here's how to make sure:

TROUBLE SHOOTING STEAM TRAPS

There are various methods for checking traps; for instance, (1) visual observation, (2) by sound, (3) by temperature measurements. Here is a description of each:

Visual Method — Visual observation of condensate discharging from the trap is the easiest, and perhaps the best way, to check its performance. No special equipment is needed, but you should know the difference between flash steam and live steam.

Flash steam is the lazy vapor that forms when hot condensate is discharged from a steam trap to the atmosphere. The presence of flash steam is natural and does not imply waste steam or trap failure. If the mixture of condensate and flash steam is being discharged several times a minute as the trap cycles, the trap is operating properly.

But suppose the steam that accompanies the condensate is not flash vapor but is live steam discharging hot, at high velocity, and (in the case of a disc-type trap) with a rapid chattering sound. Then you can assume the trap has failed and must be repaired or replaced.

Sound Method — By listening carefully to them as they operate, traps can be checked without visual observation of the condensate discharged. This method is, therefore, much more convenient when working with a closed condensate return.

Variations of the sound method range in precision from an ultrasonic testing device down through an industrial stethoscope to home-made listening tools such as a two-foot length of 3/16″ steel rod in a file handle, a welding rod, or a screwdriver.

With a little practice, the operation of the trap internals can be heard with any of these home-made devices merely by placing one end of the tool against the trap bonnet and the other end to your ear. Here's what you will hear:

Temperature Measurements — A steam trap is essentially an automatic condensate valve, the only function of which is to pass condensate and hold back steam. This definition implies the significant temperature differential that exists between the upstream and downstream sides of a properly functioning trap. Trap performance, therefore, can be checked by making temperature measurements on the pipeline immediately upstream and downstream of the trap.

Two requirements for this method are a simple contact pyrometer for making the measurements on the surface of this pipe, and a knowledge of line pressure upstream and downstream of the trap. For each steam pressure there is a corresponding steam temperature. The table shows typical pipe surface temperature readings corresponding with several operating pressures.

Let's assume the upstream pressure in the piping system is 150 psig and the pressure downstream of the trap is 15 psig. An upstream temperature measurement with the pyrometer is 335°F and a downstream reading is 225°F. (File or wire-brush the pipe at points of measurement to provide good contacts for the tip of the pyrometer.)

The table shows that for an upstream pressure of 150 psig, a pyrometer reading between 329°F and 348°F should be obtained. And for a downstream pressure of 15 psig, a pyrometer reading of between 225°F and 238°F is desirable. We can conclude, therefore, that the trap is functioning properly.

Now let's assume the same pressures, but a pyrometer reading of 335°F upstream and 300°F downstream of the trap. The insufficient spread between the two temperatures indicates that live steam is passing into the condensate return line. The trap has failed open. Repair or replace it.

In still another example, suppose the pyrometer readings are 210°F on both sides of the trap. That's OK downstream where we know pressure is 15 psig. But it's too low for a reading upstream where we know we have 150 psig in the line. There is probably a restriction in the line that is reducing the pressure to the trap. A clogged strainer may be the culprit so blow it down before looking any further for the problem.

Although the foregoing examples deal with a closed return system, the temperature measurement method can also be used to check traps discharging to the atmosphere. In this situation of course, the downstream pressure is always atmospheric.

Use All Methods — None of the methods described above provides a "cure-all" for all trap trouble shooting situations. The best results often can be achieved by using a combination of checking methods or by using one method to cross-check the indications provided by another.

CHAPTER 8

CONDENSATE REMOVAL DEVICES

RICHARD G. KRUEGER
Flexitallic Gasket Co., Inc.
Camden, NJ

INTRODUCTION

A condensate removal device is required to expel from the working vapor system all condensate, air, oxygen, CO_2 and non-condensable gases with a minimum loss of working vapor. Therefore, it is desirable to have all condensate removal devices meet the following criteria:

- Remove all condensate and undesirable elements from the system as quickly as possible with a minimum loss of working vapor (steam).
- Achieve the longest possible service life before repair or replacement is required.
- Include a quick and accurate means of checking performance.
- Be suitable for the operating conditions encountered i.e. rapid response to variations in pressure, temperature and condensate load.
- Be unaffected by water hammer and freeze-proof.
- Provide fail safe operation.
- Attain these objectives at minimum cost.

To remove condensate from steam systems, a variety of devices have been used, including traps, drilled gate valves, slotted globe valves, etc. Each of them offers advantages under a certain set of circumstances, but may prove costly in terms of steam losses and maintenance time required under other conditions. Erosion and accumulation of scale particulate are typical problems.

DRAIN ORIFICE

The Flexitallic Drain Orifice meets all the criteria for efficient condensate removal. It is designed to continuously remove all condensate as it is produced with a stable, nominally small, steam loss (normally less than 1.0 lbs/hr) (Figure 8.1).

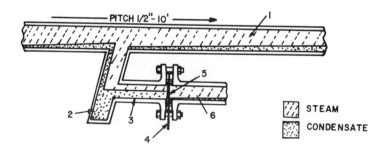

Figure 8.1 Operational drawing — all valves, etc., missing.

The Drain Orifice, when properly used in a specific application, has resulted in major reductions in energy losses and maintenance costs while significantly improving steam system reliability and performance.

What It Is

The Drain Orifice is an engineered assembly that is available in a flange-type assembly (*) (see Figure 8.2) for use in ANSI B16.5 flanges from ½″ through 1″ nominal pipe sizes up to 2500 psig pressure rating. It is also available as a union-type assembly (see Figure 8.3) for ½″, ¾″ and 1″ NPS threaded or socket weld end connections. Each assembly consists of three simple parts:

1. Spiral-wound inlet gasket with an integral fine mesh filter that serves as a secondary protection to eliminate the possibility of clogging and reduce scale erosion. A Y-Strainer installed upstream of the Drain Orifice is required as primary protection on all installations where the orifice diameter is 0.125″ or smaller.
2. Circular stainless steel plate with an orifice properly sized for operating conditions.
3. Spiral-wound outlet gasket.

The Flexitallic Drain Orifice is covered by U.S. Patent Number 3,977,895 and foreign patents and Patents Pending.

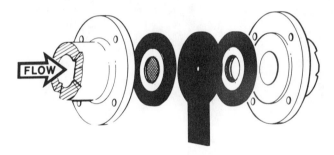

Figure 8.2 Drain Orifice (flange type).

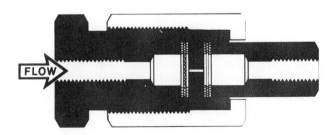

Figure 8.3 Drain Orifice union (threaded or socket weld connections).

---

*Other sizes and ratings are available.

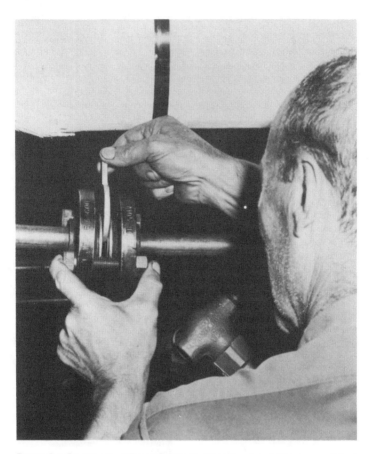

Figure 8.4 Placement of Drain Orifice in drain line is quickly accomplished without complete disassembly of flanged joint. Orifice plate is about to be placed. Entire assembly including orifice will then be insulated.

How It Works

As shown in Figure 8.1, the steam main (1) loses heat through radiation, etc., causing condensate to form. This accumulates at the bottom of line and runs to the right due to the pitch of the pipe and gravity, until it spills into the drip leg (2) where it accumulates until it reaches a level sufficient to enter the drain line (3) and reaches equilibrium at a height above the middle of the orifice plate (4) and orifice (5). The balance of the steam main (1) drip leg (2) and drain line (3) are steam filled. This highly complex flow may be visualized with the help of the following idealization: Consider that only water flowed through the system due to gravity. If the orifice had a safety factor of 1.5, then only two-thirds of the orifice would be

Figure 8.5 A Federal Incentive Award of $25,000 presented to Lawrence L. Guzick, a civilian employee of the U.S. Navy, by President Carter on May 18, 1977 for his invention of the Drain Orifice. The U.S. Navy has replaced steam traps with the Drain Orifice on over 100 ships in the Fleet. The reported annual fuel saving is in excess of $10.8 million plus $500,000 in maintenance costs.

filled and the equilibrium level would be slightly above the orifice center line. In actual conditions only a small portion of steam (by weight) passes with the condensate. Since the specific volume of steam is large and the acceleration through the orifice causes turbulence, the steam and water are thoroughly mixed.

Thus, with steam at saturation temperature, the orifice (5) passes steam at the rate shown in Graphs 8.3 and 8.4 when no condensate is present at the orifice. At operating state, it passes full condensate production and steam at approximately

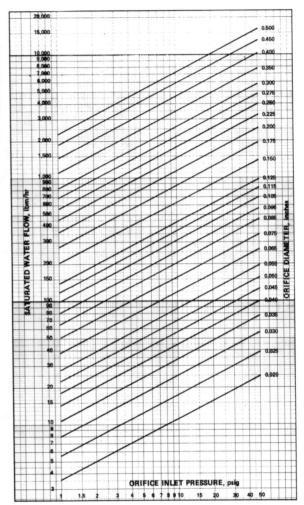

Orifice Capacity for Saturated Water
1 to 50 psig inlet pressure, zero back pressure with orifice size not exceeding:

Pipe Schedule	NPS	
	1/2"	3/4"
40	.155"D	.205"D
80	.135"D	.185"D
160	.115"D	.150"D
XX Strong	.065"D	.110"D

For correction factors for larger orifice sizes and/or greater back pressures, see Section 5.

Graph 8.1 Water flow, 1 to 50 psig.

13% of the rate shown in Graphs 8.3 and 8.4, and has the capacity to pass an additional weight of condensate equal to 50% of the normal condensate rate for that pipe/pressure/temperature rating. This is a safety factor of 1.5.

How to Determine Orifice Size

The key factor for successful performance is properly determining the orifice diameter, thereby achieving the maximum energy-saving advantages of the Drain Orifice. A properly sized orifice will achieve the objective of removing all condensate with a minimum steam loss.

176

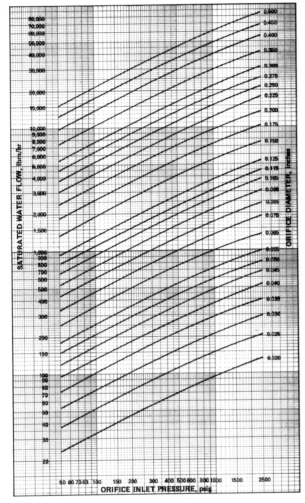

Orifice Capacity for Saturated Water
50 to 2500 psig inlet pressure, zero back pressure with orifice size not exceeding:

Pipe	NPS	
Schedule	1/2''	3/4''
40	.155''D	.205''D
80	.135''D	.185''D
160	.115''D	.150''D
XX Strong	.065''D	.110''D

For correction factors for larger orifice sizes and/or greater back pressures, see Section 5.

Graph 8.2 Water flow, 50 to 2500 psig.

To determine the proper size of the orifice, the following calculation should be used:

$$\text{Minimum Orifice Required} = \sqrt{\left[\frac{W_{wn} \times SF}{W_{wc} \times C_b}\right] \times .0004} \qquad (1)$$

177

Orifice Capacity for Saturated Steam
1 to 50 psig inlet pressure, zero back pressure with orifice size not exceeding:

Pipe	NPS	
Schedule	1/2"	3/4"
40	.155"D	.205"D
80	.135"D	.185"D
160	.115"D	.150"D
XX Strong	.065"D	.110"D

Back pressure correction is not required where r = 0.58 or less in following equation:

$$r = \frac{\text{Back Pressure (psia)}}{\text{Inlet Pressure (psia)}}$$

For correction factors for larger orifices and/or back pressure, see Section 5.

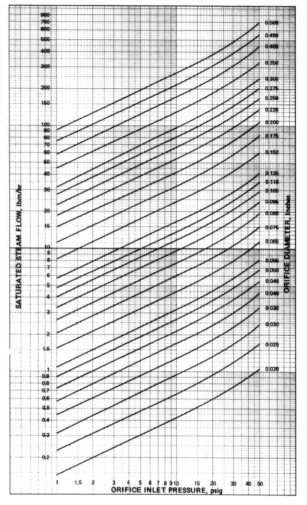

Graph 8.3 Steam flow, 1 to 50 psig.

where:

W_{wn} = condensate load, lbs/hr (calculated)

SF = safety factor — to allow for load and condition changes

W_{wc} = saturated condensate capacity of 0.020" orifice at pressure (psig)

C_b = corrections for back pressure and elevation

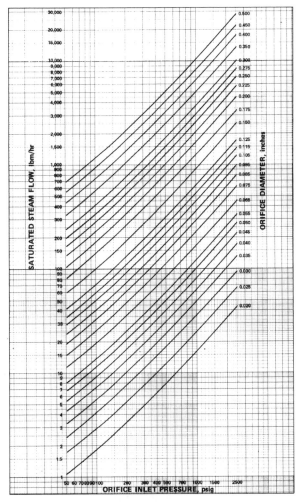

Graph 8.4 Steam flow, 50 to 2500 psig.

Orifice Capacity for Saturated Steam
50 to 2500 psig inlet pressure, zero back pressure with orifice size not exceeding:

Pipe Schedule	NPS	
	1/2"	3/4"
40	.155"D	.205"D
80	.135"D	.185"D
160	.115"D	.150"D
XX Strong	.065"D	.110"D

Back pressure correction is not required where r = 0.58 or less in following equation:

$$r = \frac{\text{Back Pressure (psia)}}{\text{Inlet Pressure (psia)}}$$

For correction factors for larger orifices and/or back pressure, see Section 5.

$$C_b = \sqrt{1 - \left[\frac{P_b + (H \times 0.5)}{P_i} \right]} \qquad (2)$$

where:

P_b = back pressure (psig)
P_i = inlet pressure (psig)
H = left head (ft)

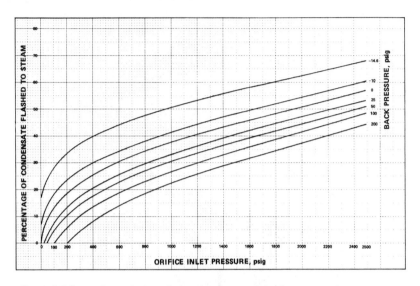

Graph 8.5 Steam formed when discharging saturated condensate to a lower pressure.

ADVANTAGES

Energy Saving Advantages

1. Offers low-initial steam loss since it is a factor of design.
2. Maintains low rate of steam loss throughout entire service life with negligible orifice erosion.
3. Eliminates possibility of large steam losses — cannot fail open.
4. Fuel savings of a significant magnitude achieved in every installation in which it has been used properly.

Maintenance Advantages

1. Provides minimum service life of more than ten years without failure or replacement.
2. Offers maintenance-free performance since there are no moving parts to wear or require replacement.

Operating Advantages

1. Accommodates normal fluctuating loads with high efficiency at constant inlet pressure.
2. Accommodates widely fluctuating loads with high efficiency if the inlet pressure is under control.
3. Unaffected by pressure surges, temperature excursions or freezing.
4. Response to load fluctuations is instantaneous — actually continuous.
5. Steam system section not in use is maintained at near operating temperature due to constant flow characteristics.

Table 8.1. Condensate Production Rate, Operating Conditions
(Pounds per hour, per linear foot of bare iron or steel pipe).

Part 1

Gage Pressure psig	NOMINAL PIPE SIZE (NPS) Inches										
	½	¾	1	1¼	1½	2	2½	3	3½	4	5
0	.080	.098	.121	.149	.169	.208	.247	.297	.336	.374	.455
1	.083	.102	.125	.155	.175	.215	.256	.307	.348	.388	.472
2	.086	.105	.129	.160	.181	.222	.265	.318	.359	.400	.487
3	.088	.108	.133	.165	.186	.229	.273	.327	.370	.413	.502
4	.091	.111	.137	.169	.192	.235	.281	.337	.381	.425	.517
5	.093	.114	.141	.174	.197	.242	.288	.346	.391	.436	.531
10	.105	.128	.158	.195	.221	.271	.324	.388	.439	.490	.596
15	.115	.141	.173	.214	.242	.298	.355	.426	.482	.538	.655
20	.124	.152	.187	.231	.262	.322	.384	.460	.521	.581	.708
25	.132	.162	.200	.247	.280	.344	.410	.492	.557	.621	.756
50	.168	.206	.253	.314	.355	.437	.521	.626	.708	.790	.963
75	.197	.242	.297	.368	.417	.513	.612	.735	.832	.929	1.132
100	.222	.273	.335	.416	.471	.579	.692	.831	.940	1.050	1.280
125	.245	.301	.370	.459	.520	.640	.764	.917	1.039	1.160	1.414
150	.266	.327	.402	.499	.565	.695	.831	.998	1.130	1.262	1.539
200	.305	.374	.461	.572	.648	.798	.953	1.146	1.298	1.449	1.768
250	.340	.418	.515	.639	.724	.892	1.066	1.281	1.452	1.622	1.978
300	.373	.459	.565	.702	.796	.980	1.172	1.409	1.597	1.784	2.177
350	.405	.498	.614	.762	.865	1.065	1.274	1.531	1.736	1.939	2.367
400	.435	.531	.660	.821	.931	1.147	1.371	1.649	1.869	2.089	2.550
450	.465	.573	.706	.877	.995	1.226	1.467	1.764	2.000	2.235	2.729
500	.495	.609	.751	.933	1.058	1.304	1.560	1.877	2.128	2.378	2.904
550	.523	.644	.794	.987	1.120	1.381	1.652	1.988	2.254	2.519	3.077
600	.552	.679	.838	1.042	1.182	1.457	1.744	2.098	2.379	2.659	3.248
700	.608	.749	.924	1.149	1.304	1.608	1.925	2.317	2.627	2.937	3.588
800	.664	.818	1.010	1.256	1.426	1.758	2.105	2.534	2.875	3.214	3.928
900	.721	.888	1.096	1.364	1.548	1.910	2.287	2.754	3.124	3.493	4.270
1000	.778	.959	1.184	1.473	1.673	2.064	2.472	2.977	3.378	3.777	4.678
1100	.837	1.031	1.274	1.586	1.800	2.222	2.662	3.206	3.638	4.068	4.974
1200	.897	1.106	1.366	1.701	1.931	2.384	2.856	3.441	3.904	4.367	5.341
1300	.959	1.183	1.461	1.820	2.067	2.552	3.057	3.683	4.180	4.675	5.719
1400	1.024	1.263	1.561	1.944	2.208	2.726	3.267	3.936	4.468	4.997	6.114
1500	1.092	1.347	1.665	2.074	2.355	2.909	3.486	4.201	4.768	5.334	6.526
1600	1.163	1.435	1.774	2.210	2.510	3.101	3.716	4.479	5.084	5.688	6.960
1700	1.239	1.528	1.889	2.354	2.674	3.304	3.960	4.773	5.419	6.062	7.419
1800	1.319	1.628	2.013	2.508	2.849	3.520	4.220	5.087	5.776	6.462	7.909
1900	1.406	1.735	2.145	2.673	3.037	3.753	4.499	5.424	6.159	6.891	8.435
2000	1.500	1.851	2.289	2.853	3.242	4.006	4.803	5.791	6.575	7.357	9.007
2100	1.603	1.978	2.447	3.049	3.466	4.283	5.136	6.193	7.032	7.869	9.633
2200	1.716	2.119	2.621	3.267	3.713	4.589	5.503	6.636	7.536	8.433	10.32
2300	1.844	2.276	2.816	3.510	3.990	4.931	5.914	7.133	8.100	9.065	11.10
2400	1.987	2.453	3.035	3.785	4.301	5.317	6.378	7.692	8.736	9.776	11.97
2500	2.159	2.665	3.298	4.112	4.674	5.779	6.931	8.361	9.495	10.83	13.02

Part 2

Gage Pressure psig	NOMINAL PIPE SIZE (NPS) Inches									
	6	8	10	12	14	16	18	20	22	24
0	.536	.685	.841	.987	1.077	1.220	1.363	1.505	1.646	1.786
1	.555	.709	.871	1.022	1.115	1.264	1.412	1.559	1.705	1.805
2	.573	.733	.900	1.056	1.153	1.306	1.459	1.611	1.762	1.912
3	.591	.756	.928	1.009	1.188	1.347	1.504	1.661	1.816	1.971
4	.608	.778	.955	1.121	1.223	1.386	1.548	1.709	1.869	2.029
5	.625	.799	.981	1.151	1.257	1.424	1.591	1.756	1.921	2.085
10	.702	.897	1.102	1.293	1.412	1.600	1.787	1.973	2.159	2.343
15	.770	.985	1.210	1.420	1.550	1.757	1.962	2.167	2.371	2.573
20	.832	1.065	1.308	1.535	1.676	1.900	2.122	2.344	2.564	2.783
25	.890	1.139	1.399	1.642	1.793	2.033	2.270	2.507	2.743	2.977
50	1.133	1.451	1.783	2.093	2.285	2.592	2.896	3.198	3.499	3.799
75	1.332	1.706	2.098	2.463	2.690	3.051	3.409	3.766	4.121	4.474
100	1.507	1.930	2.374	2.788	3.045	3.454	3.860	4.265	4.668	5.068
125	1.665	2.134	2.625	3.083	3.368	3.821	4.270	4.718	5.164	5.608
150	1.813	2.323	2.858	3.358	3.668	4.162	4.652	5.140	5.627	6.110
200	2.083	2.671	3.287	3.863	4.220	4.789	5.354	5.917	6.478	7.035
250	2.332	2.991	3.682	4.328	4.728	5.367	6.001	6.633	7.262	7.888
300	2.566	3.292	4.054	4.766	5.207	5.911	6.610	7.307	8.002	8.692
350	2.791	3.581	4.411	5.186	5.667	6.434	7.195	7.955	8.711	9.464
400	3.007	3.860	4.755	5.591	6.110	6.938	7.760	8.580	9.396	10.208
450	3.219	4.132	5.091	5.987	6.543	7.430	8.312	9.190	10.07	10.94
500	3.426	4.399	5.421	6.376	6.969	7.914	8.853	9.790	10.72	11.65
550	3.630	4.662	5.746	6.758	7.387	8.390	9.386	10.38	11.37	12.36
600	3.833	4.923	6.068	7.139	7.803	8.863	9.916	10.97	12.01	13.05
700	4.235	5.441	6.708	7.893	8.628	9.802	10.97	12.13	13.29	14.44
800	4.636	5.958	7.347	8.646	9.453	10.74	12.02	13.29	14.57	15.83
900	5.041	6.480	7.992	9.406	10.29	11.69	13.08	14.67	15.85	17.23
1000	5.452	7.010	8.647	10.18	11.13	12.65	14.16	15.66	17.16	18.65
1100	5.874	7.554	9.319	10.97	12.00	13.64	15.26	16.89	18.50	20.12
1200	6.307	8.112	10.01	11.79	12.89	14.65	16.40	18.15	19.89	21.62
1300	6.754	8.689	10.72	12.63	13.81	15.70	17.57	19.45	21.31	23.17
1400	7.221	9.291	11.47	13.50	14.77	16.79	18.80	20.80	22.80	24.79
1500	7.709	9.921	12.25	14.42	15.78	17.93	20.08	22.22	24.36	26.48
1600	8.223	10.58	13.07	15.39	16.83	19.14	21.43	23.71	25.99	28.26
1700	8.765	11.28	13.93	16.41	17.95	20.41	22.85	25.29	27.72	30.14
1800	9.345	12.03	14.86	17.50	19.14	21.77	24.37	26.98	29.57	32.16
1900	9.967	12.83	15.85	18.67	20.42	23.22	26.01	28.79	31.56	34.32
2000	10.64	13.71	16.93	19.94	21.82	24.81	27.78	30.75	33.72	36.66
2100	11.39	14.66	18.11	21.34	23.34	26.55	29.73	32.91	36.08	39.24
2200	12.20	15.72	19.41	22.88	25.03	28.46	31.88	35.29	38.69	42.07
2300	13.12	16.90	20.88	24.60	26.91	30.61	34.28	37.95	41.61	45.25
2400	14.15	18.23	22.52	26.54	29.04	33.03	36.99	40.95	44.90	48.83
2500	15.38	19.82	24.49	28.86	31.58	35.92	40.23	44.54	48.84	53.11

NOTE: Table 1 is based on an ambient temperature of 80°F in still air. To correct for other ambient temperatures, the following formula may be used for approximation:

$$C = \text{Table 1 value} \times \frac{(\text{temp. saturation} - \text{temp. ambient})}{(\text{temp. saturation} - 80°F)}$$

For exact calculation, use equation (1-3) in Section 5 of this handbook.

Table 8.2. Weight of Plain End Pipe, Pounds Per Foot of Length.

NPS	SCHEDULE										DESCRIPTION		
	10	20	30	40	60	80	100	120	140	160	Standard	X-Strong	XX-Strong
⅛				0.24		0.31					0.24	0.31	
¼				0.42		0.54					0.42	0.54	
⅜				0.57		0.74					0.57	0.74	
½				0.85		1.09				1.31	0.85	1.09	1.71
¾				1.13		1.47				1.44	1.13	1.47	2.44
1				1.68		2.17				2.84	1.68	2.17	3.66
1¼				2.27		3.00				3.76	2.27	3.00	5.21
1½				2.72		3.63				4.86	2.72	3.63	6.41
2				3.65		5.02				7.46	3.65	5.02	9.03
2½				5.79		7.66				10.01	5.79	7.66	13.69
3				7.58		10.25				14.32	7.58	10.25	18.58
3½				9.11		12.50				—	9.11	12.50	—
4				10.79		14.98		19.00		22.51	10.79	14.98	27.54
5				14.62		20.78		27.04		32.96	14.62	20.78	38.55
6				18.97		28.57		36.39		45.35	18.97	28.57	53.16
8		22.36	24.70	28.55	35.64	43.39	50.94	60.41	67.76	74.69	28.55	43.39	72.42
10		28.04	34.24	40.48	54.74	64.43	77.03	89.29	104.13	115.64	40.48	54.74	104.13
12		33.38	43.77	53.52	73.15	88.63	107.32	125.49	139.67	160.27	49.56	65.42	125.49
14	30.93	45.61	54.57	63.44	85.05	106.13	130.85	150.79	170.21	189.11	54.57	72.09	
16	42.05	52.27	62.58	82.77	107.50	136.61	164.82	197.43	223.64	245.25	62.58	82.77	
18	47.39	58.94	82.15	104.67	138.17	170.96	207.96	244.14	274.22	308.50	70.59	93.45	
20	52.73	78.60	104.13	123.11	166.40	208.87	256.10	296.37	341.09	379.17	78.60	104.13	
22	58.07	86.61	114.81	—	197.41	250.81	302.88	353.61	403.61	451.06	86.61	114.81	
24	63.41	94.62	140.68	171.29	238.35	296.58	367.39	429.39	483.12	542.13	94.62	125.49	

6. Discharge is fully controlled at all condensate loads.
7. Rated capacity is at saturation temperature, either subcooling of condensate or super-heated steam conditions increase efficiency of operation.

Limitations

The Drain Orifice is not a panacea but can be used in approximately 70% of all applications now being handled by steam traps. It should not be used where the condensate load fluctuates more than half of the safety factor oversizing, unless cascaded.

Example —

Applying a safety factor of 1.5 (50% Oversizing), the load change would be restricted to ±25% of design load.

Applications

The Drain Orifice can be used successfully in the following applications:

1. Main steam supply drainage
2. Warm-up conditions
3. Boiler blowdown
4. Tubular trace heating
5. Space heaters
6. Process heaters
7. Cascaded drains
8. Heat exchangers

Example —

Heater rated 80,000 BTU/hour at 2 psig steam, 60°F entering air temperature. Heater applied at 50 psig steam, 50°F entering air temperature.

Table 8.3. Correction Factor for BTU Rated Unit Heaters Applied at Non-Standard Conditions.

h_{fg}	PRESSURE Psig	ENTERING AIR TEMPERATURE											
		−10°F	0°F	10°F	20°F	30°F	40°F	50°F	60°F	70°F	80°F	90°F	100°F
970.3	0	1.527	1.440	1.354	1.271	1.190	1.111	1.034	.9589	.8856	.8140	.7441	.6758
968.2	1	1.550	1.462	1.376	1.293	1.212	1.133	1.055	.9800	.9065	.8347	.7646	.6960
966.2	2	1.572	1.484	1.398	1.314	1.232	1.153	1.076	1.000	.9263	.8542	.7839	.7151
964.2	3	1.593	1.504	1.418	1.334	1.252	1.172	1.095	1.019	.9450	.8728	.8022	.7333
962.3	4	1.613	1.524	1.437	1.353	1.271	1.191	1.113	1.037	.9629	.8905	.8198	.7506
960.6	5	1.632	1.542	1.456	1.371	1.289	1.209	1.131	1.054	.9801	.9075	.8366	.7673
952.5	10	1.716	1.626	1.538	1.452	1.369	1.288	1.209	1.132	1.056	.9830	.9113	.8411
945.6	15	1.787	1.696	1.607	1.521	1.437	1.355	1.275	1.197	1.121	1.047	.9744	.9036
939.5	20	1.849	1.757	1.668	1.580	1.496	1.413	1.332	1.254	1.177	1.103	1.029	.9580
933.9	25	1.904	1.812	1.721	1.634	1.548	1.465	1.384	1.305	1.227	1.152	1.078	1.006
928.9	30	1.954	1.861	1.770	1.681	1.595	1.511	1.430	1.350	1.272	1.197	1.122	1.050
924.2	35	1.999	1.906	1.814	1.725	1.639	1.554	1.472	1.392	1.314	1.237	1.163	1.090
919.8	40	2.042	1.947	1.855	1.765	1.678	1.593	1.511	1.430	1.352	1.275	1.199	1.127
915.6	45	2.080	1.985	1.893	1.803	1.715	1.630	1.547	1.466	1.387	1.310	1.234	1.161
911.7	50	2.117	2.021	1.928	1.838	1.750	1.664	1.581	1.499	1.420	1.342	1.267	1.193
904.5	60	2.183	2.087	1.993	1.902	1.813	1.727	1.642	1.560	1.480	1.402	1.326	1.251
897.9	70	2.243	2.146	2.051	1.960	1.870	1.783	1.698	1.615	1.534	1.455	1.379	1.303
894.8	75	2.271	2.173	2.078	1.986	1.896	1.809	1.723	1.640	1.559	1.480	1.403	1.328
891.7	80	2.297	2.199	2.104	2.011	1.921	1.834	1.748	1.665	1.583	1.504	1.427	1.351
885.9	90	2.347	2.247	2.153	2.060	1.969	1.880	1.794	1.711	1.629	1.549	1.471	1.395
880.6	100	2.394	2.294	2.198	2.104	2.013	1.924	1.837	1.753	1.671	1.590	1.512	1.435
868.2	125	2.497	2.396	2.298	2.203	2.111	2.021	1.933	1.848	1.764	1.683	1.603	1.526
857.0	150	2.586	2.484	2.385	2.289	2.196	2.105	2.016	1.930	1.845	1.763	1.683	1.604
837.3	200	2.726	2.622	2.522	2.424	2.329	2.236	2.146	2.058	1.972	1.888	1.807	1.727
820.1	250	2.862	2.757	2.655	2.555	2.458	2.364	2.272	2.183	2.096	2.011	1.928	1.847

Solution —

$$C = \frac{80,000 \times 1.581}{911.7} = 138.7 \text{ lbs/hour} \qquad (E)$$

c) CFM Rated Heaters

$$C, \text{lbs/hour} = \frac{1.078 \times CF^* \times SCFM \times (\text{Rated Final Air Temp.} - 60)}{h_{fg} \, (\text{Applied Pressure})} \qquad (F)$$

---

* From Table 8.3

Table 8.4. Properties of Saturated Steam.

Pressure		Temperature	Specific Volume ft /lbm		Enthalpy BTU /lbm		
Gage psig	Absolute psia	Saturation F	Liquid v	Steam v	Liquid h	Latent h	Steam h
0	14.7	212.00	0.016719	26.799	180.17	970.3	1150.5
1	15.7	215.35	0.016742	25.198	183.54	968.2	1151.7
2	16.7	218.52	0.016764	23.783	186.73	966.2	1152.9
3	17.7	221.53	0.016786	22.253	189.77	964.2	1154.0
4	18.7	224.40	0.016807	21.394	192.67	962.3	1155.0
5	19.7	227.16	0.016828	20.376	195.45	960.6	1156.0
10	24.7	239.40	0.016922	16.488	207.84	952.5	1160.4
15	29.7	249.76	0.017004	13.8766	218.33	945.6	· 1163.9
20	34.7	258.78	0.017079	11.9941	227.5	939.5	1167.0
25	39.7	266.80	0.017147	10.5722	235.6	933.9	1169.6
50	64.7	297.67	0.017430	6.6826	267.3	911.7	1179.0
75	89.7	320.04	0.017657	4.9109	290.5	894.8	1185.2
100	114.7	337.89	0.01785	3.8910	309.1	880.6	1189.6
125	139.7	352.87	0.01803	3.2566	324.8	868.2	1192.9
150	164.7	365.87	0.01818	2.7562	338.6	857.0	1195.6
200	214.7	387.79	0.01847	2.13609	362.0	837.3	1199.3
250	264.7	406.03	0.01873	1.74353	381.6	820.1	1201.7
300	314.7	421.75	0.01896	1.47200	398.8	804.5	1203.3
350	364.7	435.65	0.01919	1.27287	414.2	790.0	1204.2
400	414.7	448.14	0.01940	1.12002	428.2	776.6	1204.7
450	464.7	459.52	0.01961	0.99904	441.0	763.8	1204.8
500	514.7	470.00	0.01980	0.90080	453.0	751.6	1204.5
550	564.7	479.72	0.02000	0.81938	464.1	740.0	1204.1
600	614.7	488.81	0.02019	0.75077	474.7	728.7	1203.5
700	714.7	505.40	0.02056	0.64140	494.4	707.1	1201.5
800	814.7	520.31	0.02092	0.55797	512.4	686.6	1199.0
900	914.7	533.87	0.02128	0.49213	529.1	666.8	1195.9
1000	1014.7	546.35	0.02164	0.43876	544.8	647.6	1192.4
1100	1114.7	557.93	0.02201	0.39458	559.7	628.7	1188.5
1200	1214.7	568.73	0.02237	0.35736	573.9	610.2	1184.2
1300	1314.7	578.88	0.02275	0.32554	587.6	592.0	1179.5
1400	1414.7	588.44	0.02313	0.29797	600.7	573.8	1174.6
1500	1514.7	597.50	0.02352	0.27384	613.5	555.7	1169.3
1600	1614.7	606.10	0.02393	0.25247	626.0	537.6	1163.7
1700	1714.7	614.31	0.02435	0.23339	638.3	519.5	1157.7
1800	1814.7	622.14	0.02479	0.21620	650.3	501.1	1151.3
1900	1914.7	629.65	0.02524	0.20057	662.1	482.5	1144.6
2000	2014.7	636.84	0.02572	0.18629	673.9	463.4	1137.2
2100	2114.7	643.76	0.02623	0.17314	685.5	443.8	1129.3
2200	2214.7	650.40	0.02677	0.16099	697.2	423.7	1120.9
2300	2314.7	656.82	0.02736	0.14973	708.9	402.9	1111.9
2400	2414.7	663.01	0.02799	0.13925	720.8	381.5	1102.2
2500	2514.7	668.98	0.02870	0.12925	733.6	358.1	1091.7

Example —

Heater rated at 1,420 CFM at 70°F, 90,600 BTU/hour, final air temperature 119°F. Heater applied at 20 psig saturated steam, entering air at 50°F.

Solution —

From Table 8.3 at 20 psig, h_{fg} = 939.5 BTU/lbm; at 20 psig and 50°F, CF = 1.332

$$C = \frac{1.078 \times 1.332 \times 1,420 \times (119 - 60)}{939.5} = 128.0 \text{ lbs/hour} \qquad \text{(F)}$$

Step 3) Multiply condensate load (Step 2) by a safety factor of 2.0 to reach the minimum desired orifice capacity for saturated liquid in this application. NOTE: Increase in safety factor—standard for all devices to prevent short circuiting, pressure differential, etc., common to heat transfer equipment.

Step 4) Locate orifice size which meets or exceeds required capacity (Step 3) in Graph 8.1 or 8.2.

Step 5) Use drain NPS, pressure rating (Step 1) and orifice size (Step 4) to order:

Example —

Standard BTU rated heater applied at standard conditions. Rating 150,000 BTU/hour. Drain line is ¾" NPS, 150 lb rating, zero psig back pressure.

Solution —

Step 1) Drain line is ¾" NPS, 150 lb rating

Step 2) 150,000/966.2 = 155.2 lbs/hour condensate

Step 3) 155.2 × 2.0 = 310.4 lbs/hour minimum desired capacity for orifice

Step 4) From Graph 8.1 at 2 psig, orifice required is 0.175"D

Step 5) Specify Flexitallic #¾-150 DFA Style 125P (0.125")

ALTERNATE PROCEDURE
FOR CLOSE TOLERANCE SIZING OF ORIFICE

To calculate the orifice diameter required with greater accuracy and efficiency than can be achieved using the graph sizes shown in the preceding pages, use the following procedure:

Step 1) Calculate the minimum condensate capacity required including the safety factor according to the standard procedures in this section.

Step 2) Determine the condensate capacity for an 0.020"D orifice at the proper inlet pressure from Graph 8.1 or 8.2.

Step 3) Divide the required capacity (Step 1) by the 0.020"D condensate capacity (Step 2) to determine the orifice minimum multiplier.

Step 4) Locate the multiplier in Table 8.5 that is equal to the minimum multiplier (Step 3) or the next higher multiplier. The adjacent orifice size is the orifice size required.

Example —

Standard BTU rated heater applied at standard conditions. Rating 150,000 BTU/hour. Drain line is ¾" NPS, 150 lb rating, zero psig back pressure.

Table 8.5. Multiplier Chart for Orifice Diameter Sizing.

Multiplier	Orifice ± 0.001"	Multiplier	Orifice ± 0.001"	Multiplier	Orifice ± 0.001"	Multiplier	Orifice ± 0.001"	Multiplier	Orifice ± 0.001"	Multiplier	Orifice ± 0.001"
1.00	0.0200"	4.41	0.0420"	21.85	0.0935"	51.84	0.1440"	91.20	0.1910"		
1.10	0.0210"	4.62	0.0430"	21.94	0.0937"	54.02	0.1470"	93.60	0.1935"		
1.26	0.0225"	5.40	0.0465"	23.04	0.0960"	55.87	0.1495"	96.04	0.1960"		
1.44	0.0240"	5.49	0.0469"	24.01	0.0980"	57.76	0.1520"	99.00	0.1990"		
1.56	0.0250"	6.76	0.0520"	24.75	0.0995"	59.29	0.1540"	101.00	0.2010"		
1.69	0.0260"	7.56	0.0550"	25.75	0.1015"	60.99	0.1562"	103.12	0.2031"		
1.96	0.0280"	8.85	0.0595"	27.04	0.1040"	61.62	0.1570"	104.04	0.2040"		
2.13	0.0292"	9.76	0.0625"	28.35	0.1065"	63.20	0.1590"	105.57	0.2055"		
2.40	0.0310"	10.08	0.0635"	29.86	0.1093"	64.80	0.1610"	109.20	0.2090"		
2.43	0.0312"	11.22	0.0670"	30.25	0.1100"	68.89	0.1660"	113.42	0.2130"		
2.56	0.0320"	12.25	0.0700"	30.80	0.1110"	71.82	0.1695"	119.57	0.2187"		
2.72	0.0330"	13.32	0.0730"	31.92	0.1130"	73.87	0.1719"	122.10	0.2210"		
3.06	0.0350"	14.44	0.0760"	33.64	0.1160"	74.82	0.1730"	129.96	0.2280"		
3.24	0.0360"	15.24	0.0781"	36.00	0.1200"	78.32	0.1770"	136.89	0.2340"		
3.42	0.0370"	15.40	0.0785"	39.06	0.1250"	81.00	0.1800"	137.35	0.2344"		
3.61	0.0380"	16.40	0.0810"	41.28	0.1285"	82.81	0.1820"	141.61	0.2380"		
3.80	0.0390"	16.81	0.0820"	46.24	0.1360"	85.56	0.1850"	146.41	0.2420"		
4.00	0.0400"	18.49	0.0860"	49.35	0.1405"	87.89	0.1875"	151.29	0.2460"		
4.20	0.0410"	19.80	0.0890"	49.42	0.1406"	89.30	0.1890"	156.25	0.2500"		

NOTE 1: Multipliers may also be used to calculate steam capacity of orifice.

NOTE 2: Multipliers not included in Table 5 may be calculated:

$$Multiplier = \frac{D^2}{0.0004}$$

NOTE 3: Approximate orifice diameter required may be calculated:

$$Diameter = \sqrt{Multiplier \times 0.0004}$$

Solution —

Step 1) From example in the center of page 185 (Step 3), the desired condensate capacity is 310.4 lbs/hour.

Step 2) Capacity of an 0.020"D orifice at 2 psig is 5.0 lbs/hour.

Step 3) 310.4 ÷ 5.0 = 62.08 minimum multiplier.

Step 4) From Table 8.5, the nearest (next higher) multiplier is 63.20. The adjacent column shows that an orifice diameter of 0.1590" is the proper size.

REFERENCES

1. Bulletin 474, "Design Handbook, Flexitallic Drain Orifice System," Flexitallic Gasket Company Inc., Special Products Division, Camden, NJ (1976).
2. Bulletin 776, Flexitallic Drain Orifice, Flexitallic Gasket Company Inc., Special Products Division, Camden, NJ.
3. "Phaseout of High Pressure Steam Traps," by Lawrence L. Guzick, *Naval Engineers Journal*, April 1973.
4. John C. Vaaler Award, Honors Award, Energy Saving Development Category, *Chemical Processing*, mid-November 1976.
5. Carter-Guzick Award, the Naval Observer, Vol. I, No. 18, May 19, 1977.
6. "Drop Demand on Steam Generators with Fixed-Rate Condensate Drain," by Paul Kline, Process Consultant, the Dow Chemical Company, Michigan Division, Midland, Michigan and Al Gaines, Associate Editor, *Chemical Processing*, June 1976.
7. "A Direct Approach to Large Energy and Maintenance Savings in the Refining-Petrochemical Industry," by Robert E. Beatty, Product Manager and Richard G. Krueger, Marketing Manager, Flexitallic Gasket Company Inc., Special Products Division, Camden, NJ. Paper No. 75C presented at 70th Annual Meeting, AICHE, Session 58, New York, NY on November 16, 1977.
8. Public commendation to R. Beatty, J. O'Donnel, Jr., Wall O'Donnel Gases dated June 7, 1978.

CHAPTER 9

THE PLATE HEAT EXCHANGER

A. COOPER
Technical Development
APV Company, Inc.
Tonawanda, NY

INTRODUCTION

While the original idea for the plate heat exchanger was patented in the latter half of the past century, the first commercially successful design was introduced in 1923 by Dr. Richard Seligman, founder of APV. Initially, a number of cast gun-metal plates were enclosed within a frame in a manner quite similar to a filter press. The early 1930s, however, saw the introduction of plates pressed in thin gauge stainless steel. While the basic design remains unchanged, continual refinements have boosted operating pressures from about 15 psi to 300 psi in current machines.

The (PHE) plate heat exchanger consists of a frame in which closely spaced metal plates are clamped between a head and follower. The plates have corner ports and are sealed by gaskets around the ports and along the plate edges. A double seal forms pockets open to atmosphere to prevent mixing of product and service liquids in the rare event of leakage past a gasket. A typical plate is shown (Figure 9.1).

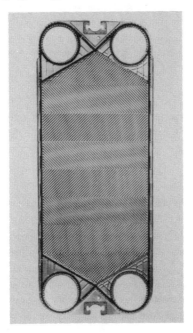

Figure 9.1

Comparative Plate Arrangements

Clarification of PHE arrangements with those for a tubular exchanger is detailed in Table 9.1. Essentially, the number of passes on the tube side of a tubular unit can be compared with the number of passes on a plate heat exchanger. The number of tubes per pass also can be equated with the number of passages per pass for the PHE. However, the comparison with the shell side usually is more difficult since with a PHE, the total number of passages available for the flow of one fluid must equal those available for the other fluid to within ±1. The number of cross passes on a shell, however, can be related to the number of plate passes and since the number of passages/pass for a plate is an indication of the flow area, this can be equated to the shell diameter. This is not a perfect comparison, but it does show the relative parameters for each exchanger.

Table 9.1. Pass Arrangement Comparison: Plate vs Tubular.

	SHELL AND TUBE	PLATE EQUIVALENT	
Tube Side	No. of Passes	No. of Passes	Side One
	Number of tubes/pass	Number of passages/pass	
Shell Side	No. of cross-passes (No. of baffles +1)	No. of passes	Side Two
	Shell diameter	Number of passages/pass	

An important, exclusive feature of the plate heat exchanger is that by the use of special connector plates it is possible to provide connections for alternate fluids so that a number of duties can be done in the same frame.

Plate Construction

Plates are pressed from stainless steel, titanium, Hastelloy C, Incoloy 825, nickel 200, Monel 400, aluminum brass, tantalum, or any material ductile enough to be formed into a pressing. A special trough and dimple pattern strengthens the plates, increases the effective heat transfer area and produces turbulence in the liquid flow between plates. Plates are pressed in material between 0.024 in^2 and 0.048 in^2 and the degree of mechanical loading is important. The most severe case occurs when one process liquid is operating at the highest working pressure and the other at zero pressure. The maximum pressure differential is applied across the plate and results in a considerable unbalanced load that tends to close the typical 0.1–0.2 in^2 gap. It is essential that some form of interplate support is provided to hold the gap and this is done by two alternative trough forms.

One method is to press pips into a plate with deep corrugations to provide contact points for about every 1 to 3 in^2 of heat transfer surface. Another is the crisscross plate of relatively shallow troughs with support maintained by corrugation/corrugation contact. Alternate plates are arranged so that corrugations cross to provide a contact point for every 0.2 to 1 in^2 of area. The plate then can handle large differential pressures and the cross pattern forms a tortuous path that promotes substantial liquid turbulence and a very high heat transfer coefficient. The net result is high rates with moderate pressure drop.

Plates are available with effective heat transfer areas from 0.28 ft^2 to 23.0 ft^2 and up to 700 of any one size can be contained in a large frame. The largest unit thereby provides 16,100 ft^2 of surface area. Flow ports and associated pipework are sized in proportion to the plate area and control the maximum liquid throughput.

Gasket Materials

As detailed in Table 9.2, various gasket elastomers are available which have chemical and temperature resistance coupled with good sealing properties. The temperatures shown are maximum so possible simultaneous chemical action must be taken into account.

Table 9.2. Gasket Materials — Operating Temperatures and Applications.

Gasket Material	Approx. Maximum Operating Temp.		Application
	°F	°C	
Medium Nitrile	275	135	Resistant to fatty materials.
EPDM	300	150	High temperature resistance for a wide range of chemicals.
· Resin cured butyl	300	150	Aldehydes, ketone and some esters.
Fluorocarbon rubber base	350	177	Mineral oils, fuels, vegetable and animal oils.
Compressed asbestos fiber	500	260	Organic solvents such as chlorinate hydrocarbons.

To ensure resistance to chemicals and temperatures as high as 500°F, gaskets of compressed asbestos fiber have been developed. Since this material has virtually no elasticity compared to rubber gaskets, the gasket groove in the plate and frame are designed to withstand the much greater compression force required to obtain a pressure seal. These gaskets are made from two laminates of material and since they tend to adhere to the adjacent metal, a special release agent is used to simplify opening of the machine while still maintaining the seal in good condition.

Thermal Performance

To exemplify typical thermal data, this report covers the performance of PHEs with turbulent liquid/turbulent liquid flow. Heat transfer can best be described by a Dittus Boelter type equation:

$$Nu = (A) \ (Re)^n \ (Pr)^m \ \left(\frac{\mu}{\mu_w} \right)^x$$

Reported values of the constant and exponents are:

A = .15 to .40
n = .65 to .85
m = .30 to .45
x = .05 to .20

where
$$Nu = \frac{hd}{k} \quad Re = \frac{Vd\rho}{\mu} \quad Pr = \frac{Cp\mu}{k}$$

d is the equivalent diameter defined in the case of the plate heat exchanger as 2X the mean gap.

Typical velocities in plate heat exchangers for waterlike fluids in turbulent flow are 1 to 3 ft/s but true velocities in certain regions will be higher by a factor of up to 4 due to the effect of the corrugations. All heat transfer and pressure drop relationships are, however, based on either a velocity calculated from the average plate gap or on the flow rate per passage.

Figure 9.2 illustrates the effect of velocity for water at 60°F on heat transfer coefficients. This graph also plots pressure drop against velocity under the same conditions. The film coefficients are very high and can be obtained for a moderate pressure drop.

One particularly important feature of the PHE is that the turbulence induced by the troughs reduces the Reynolds number at which the flow becomes laminar. If the characteristic length dimension in the Reynolds number (Re) is taken as twice the average gap between plates, the Re number at which the flow becomes laminar varies from about 100 to 400 according to the type of plate.

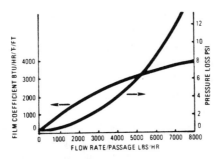

Figure 9.2 Performance details: Series HX Paraflow PHE. (Courtesy of APV Company, Inc.)

To achieve these high coefficients, it is necessary to expend energy. With the plate unit, the friction factors normally encountered are in the range of 10 to 400 times those inside a tube for the same Reynolds number. However, nominal velocities are low and plate lengths do not exceed 7.5 ft so that the term $(V^2)L/(2g)$ in the pressure drop equation is very much smaller than one would normally encounter in tubulars. In addition, single pass operation will achieve many duties so that the pressure drop is efficiently used and not wasted on losses due to flow direction changes.

The friction factor is correlated with an equation:

$$\Delta p = \frac{f.L\rho V^2}{2g.d}$$

$$f = \frac{B}{(Re)^y}$$

where y varies from 0.1 to 0.4 according to the plate and B is a constant characteristic of the plate. If the overall heat transfer equation $H = US\Delta T$ is used to calculate the heat duty, it is necessary to know the overall coefficient U, the surface area S and the mean temperature difference ΔT.

The overall coefficient U can be calculated from

$$\frac{1}{U} = r_{fh} + r_{fc} + r_w + r_{dh} + r_{dc}$$

The values of r_{fh} and r_{fc} (the film resistances for the hot and cold fluids respectively) can be calculated from the Dittus Boelter equations previously described and the wall metal resistance, r_w, can be calculated from the average metal thickness and thermal conductivity. The fouling resistances of the hot and cold fluids, r_{dh} and r_{dc}, are often based on experience but a more detailed discussion of this will be presented later.

The value taken for S is the developed area after pressing. That is the total area available for heat transfer and, due to the corrugations, will be greater than the projected area of the plate, i.e., 1.81 ft^2 vs 1.45 ft^2 for an APV HX plate.

The value of ΔT is calculated from the logarithmic mean temperature difference multiplied by a correction factor. With single pass operation, this factor is about 1 except for plate packs of less than 20 when the end effect has a significant bearing on the calculation. This is due to the fact that the passage at either end of the plate pack only transfers heat from one side and therefore the head load is reduced.

When the plate unit is arranged for multiple pass use, a further correction factor must be applied. Even when two passes are countercurrent to two other passes, at least one passage must experience co-current flow. This correction factor is shown in Figure 9.3 against a number of heat transfer units (HTU = temperature rise of the process fluid divided by the mean temperature difference). As indicated, whenever unequal passes are used, the correction factor calls for considerable increase in area. This is particularly important when unequal flow conditions are handled. If high and low flow rates are to be handled, the necessary velocities must be maintained with the low fluid flow rate by using an increased number of passes. Although the plate unit is most efficient when the flow ratio between two fluids is in the range of 0.7–1.4, other ratios can be handled with unequal passes. This is done, however, at the expense of the LMTD factor.

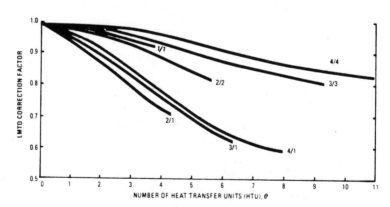

Figure 9.3 LMTD correction factor.

Conclusion

It has been shown, therefore, that the PHE is a relatively simple machine on which to carry out a thermal design. Unlike the shell side of tubular exchangers where predicting performance depends on baffle/wall leakage, baffle/tube leakage and leakage around the bundle, it is not possible to have bypass streams on a PHE. The only major problem is that the pressure loss through the ports can cause unequal distribution in the plate pack. This is overcome by limiting the port velocity and by using a port pressure loss correlation in the design to allow for the effect of unequal distribution.

The flow in a plate also is far more uniform than on the shell side. Furthermore, there are no problems over calculation of heat transfer in the window, across the bundle or of allowing for dead spots as is the case with tubular exchangers. As a result, the prediction of performance is simple and very reliable once the initial correlations have been established.

DUTIES OTHER THAN TURBULENT LIQUID FLOW

Liquid/Liquid

Over many years two manufacturers have built up considerable experience in the design and use of plate heat exchangers for process applications that fall outside the normal turbulent liquid flow that is common in chemical operations. The PHE, for example, can be used in laminar flow duties, for the evaporation of fluids with relatively high viscosities, for cooling various gases, and for condensing applications where pressure drop parameters are not overly restrictive.

Condensing

One of the most important heat transfer processes within the CPI is the condensation of vapors — a duty which often is carried out on the shell side of a tubular exchanger but is entirely feasible in the plate type unit. Generally speaking the determining factor is pressure drop.

For those condensing duties where permissible pressure loss is less than one psi, there is no doubt but that the tubular unit is most efficient. Under such pressure drop conditions, only a portion of the length of a plate would be used and substantial surface area would be wasted. However, when less restrictive pressure drops are available the plate heat exchanger becomes an excellent condenser since very high heat transfer coefficients are obtained and the condensation can be carried out in a single pass across the plate.

Pressure Drop of Condensing Vapors

The pressure drop of condensing steam in the passages of plate heat exchangers has been investigated experimentally for a series of different plates. As indicated in Figure 9.4 which provides data for a typical unit, the drop obtained is plotted against steam flow rate per passage for a number of inlet steam pressures.

It is interesting to note that for a set steam flow rate and a given duty the steam

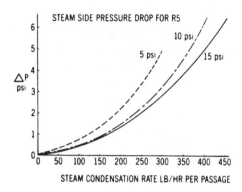

Figure 9.4 Steam condensation rate lb/hr per passage.

pressure drop is higher when the liquid and steam are in countercurrent rather than cocurrent flow. This is due to differences in temperature profile.

From Figure 9.5 it can be seen that for equal duties and flows, the temperature difference for countercurrent flow is lower at the steam inlet than at the outlet with most of the steam condensation taking place in the lower half of the plate. The reverse holds true for cocurrent flow. In this case, most of the steam condenses in the top half of the plate, the mean vapor velocity is lower and a reduction in pressure drop of between 10–40% occurs. This difference in pressure drop becomes lower for duties where the final approach temperature between the steam and process fluid becomes larger.

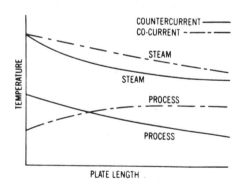

Figure 9.5 Temperature profile during condensation of steam.

The pressure drop of condensing steam, therefore, is a function of steam flow rate, pressure and temperature difference. Since the steam pressure drop affects the saturation temperature of the steam, the mean temperature difference in turn becomes a function of steam pressure drop. This is particularly important when vacuum steam is being used since small changes in steam pressure can give significant changes in the temperature at which the steam condenses.

By using a computer program and a Martinelli Lockhart type approach to the problem, APV has correlated the pressure loss to a high degree of accuracy. Table 9.3 cites a typical performance of steam heated Series R4 APV Paraflow. From this experimental run during which the exchanger was equipped with only a small number of plates, it can be seen that for a 4–5 psi pressure drop and above, the plate is completely used. Below that figure, however, there is insufficient pressure drop available to fully use the entire plate and part of the surface therefore is flooded to reduce the pressure loss. At only one psi allowable pressure drop, only 60% of the plate is used for heat transfer which is not particularly economic.

Table 9.3. Steam Heating in a PHE.

Water flow rate 16,000 lbs/hr				
Inlet water temperature 216°F				
Inlet steam temperature 250°F				
Total number of plates - 7				
Available pressure loss psi	1	2	4	6
Total duty BTU/hr	207,000	256,000	320,000	333,000
Fraction of plate flooded	40	30	4	0
Effective overall heat transfer coefficient, clean	445	520	725	770
Pressure loss psi	1	2	4	4.5

The example however, well illustrates the application of a plate heat exchanger to condensing duties. If sufficient pressure loss is available, then the plate type unit is a good condenser. The overall coefficient of 770 BTU/ft^{2}hr°F for 4–5 psi pressure loss is much higher than a coefficient of 450–500 BTU/ft^{2}hr°F which could be expected in a tubular exchanger of this type of duty. However, the tubular design would, for shell side condensation, be less dependent on available pressure loss and for a one psi drop, a 450–500 BTU/ft^{2}hr°F overall coefficient still could be obtained. With the plate, the calculated coefficient at this pressure is 746 BTU/ft^{2}hr°F but the effective coefficient based on total area is only 60% of that figure or 445 BTU/ft^{2}hr°F.

Gas Cooling

Plate heat exchangers also are used for gas cooling with units in service for cooling moist air, hydrogen and chlorine. The problems are similar to those of steam heating since the gas velocity changes along the length of the plate due either to condensation or pressure fluctuations. Designs usually are restricted by pressure drop so machines with low pressure drop plates are recommended. A typical allowable pressure loss would be 0.5 psi with rather low gas velocities giving overall heat transfer coefficients in the region of 50 BTU/ft² hr°F.

Evaporating

The plate heat exchanger also can be used for evaporation of highly viscous fluids when the evaporation occurs in the plate or as a flash unit when the liquid flashes after leaving the plate. Applications generally have been restricted to the soap and food industries. The advantage of these units is their ability to concentrate viscous fluids of up to 50 poise.

Laminar Flow

One other field suitable for the plate heat exchanger is that of laminar flow heat transfer. It has been previously pointed out that the PHE can save surface by handling fairly viscous fluids in turbulent flow because the critical Reynolds number is low. Once the viscosity exceeds 20–50 cP, however, most plate heat exchanger designs fall into the viscous flow range. Considering only Newtonian fluids since most chemical duties fall into this category, in laminar ducted flow the flow can be said to be one of three types: 1) fully developed velocity and temperature profiles (i.e. the limiting Nusselt case), 2) fully developed velocity profile with developing temperature profile (i.e. the thermal entrance region), or 3) the simultaneous development of the velocity and temperature profiles.

The first type is of interest only when considering fluids of low Prandtl number and this does not usually exist with normal plate heat exchanger applications. The third is relevant only for fluids such as gases which have a Prandtl number of about 1. Therefore, consider type two.

As a rough guide for plate heat exchangers, the rate of the hydrodynamic entrance length l_{HYD}, to the corresponding thermal entrance length l_{TH}, is given by

$$\frac{l_{TH}}{l_{HYD}} = 1.7Pr$$

Correlations for heat transfer and pressure drop in laminar flow are:

Laminar Flow Heat Transfer

$$Nu = c\left(\frac{Re.Pr.d}{L}\right)^{1/3}\left(\frac{\mu}{\mu_w}\right)^{n}$$

where   Nu   = Nusselt number, hd/k

           Re   = Reynolds number, $(vd\rho)/\mu$

           Pr   = Prandtl number, $Cp.\mu/k$

           L   = Nominal plate length

           d   = Equivalent diameter, (2X average plate gap)

           $(\mu/\mu_w)^n$ = Sieder Tate correction factor

           c   = Constant for each plate (usually in the range 1.86 to 4.50)

           n   = Index varying from 0.1–0.2 depending upon plate type

Pressure Drop

For pressure loss in a plate, the friction factor can be taken as

$$f = \frac{a}{Re}$$

when "a" is a constant characteristic of the plate.

It can be seen that for heat transfer, the plate heat exchanger is ideal because the value of "d" is small and the film coefficients are proportional to $d^{-2/3}$. Unfortunately, however, the pressure loss is proportional to $(d)^{-4}$, and pressure drop is sacrificed to achieve the heat transfer.

From these correlations, it is possible to calculate the film heat coefficient and the pressure loss for laminar flow. This coefficient combined with the metal coefficient and the calculated coefficient for the service fluid together with the fouling resistance then are used to produce the overall coefficient. As with turbulent flow, an allowance has to be made to the LMTD to allow for either end effect correction for small plate packs and/or concurrency caused by having concurrent flow in some passages. This is particularly important for laminar flow since these exchangers usually have more than one pass.

Conclusions

To summarize, it can be stated that the plate heat exchanger can be used effectively for many different types of duty other than the common liquid/liquid duties in turbulent flow. Many PHEs are in use for condensation gas cooling and laminar flow and are proving more economical to operate than other types of exchangers.

THE PROBLEM OF FOULING

The Fouling Factor

In view of its complexity, variability and the need to carry out experimental work on a long term basis under actual operating conditions, fouling remains a somewhat neglected issue among the technical aspects of heat transfer. Still, the importance of carefully predicting fouling resistance in both tubular and plate heat exchanger calculations cannot be overstressed. This is well illustrated by the following tables.

Table 9.4. Typical Water/Water Tubular Design.
(Clean overall coefficient 500 BTU/hr ft^2 °F.)

Fouling Resistance (BTU/hr ft^2 °F)$^{-1}$	Dirty Coefficient BTU/hr ft^2 °F	% Extra Surface Required
.0002	455	10
.0005	400	25
.001	333	50
.002	250	100

Table 9.5. R405 Paraflow, Water/Water Duty.
(Overall coefficient 1000 BTU/hr ft^2 °F.)
(Single pass — pressure loss 9 psi.)

Fouling Resistance (BTU/hr ft^2 °F)$^{-1}$	Dirty Coefficient BTU/hr ft^2 °F	% Extra Surface Required
.0002	833	20
.0005	666	50
.001	500	100
.002	333	200

Note that for a typical water/water duty in a plate heat exchanger, it would be necessary to double the size of the unit if a fouling factor of .0005 was used on each side of the plate (i.e. a total fouling of .001).

Although fouling is of great importance, there is relatively little accurate data available and the rather conservative figures quoted in Kern (Process Heat Transfer) are used all too frequently. It also may be said that many of the high fouling resistances quoted have been obtained from poorly operated plants. If a clean exchanger, for example, is started and run at the designed inlet water temperature, it will exceed its duty. To overcome this, plant personnel tends to turn down the cooling water flow rate and thereby reduce turbulence in the exchanger. This encourages fouling and even though the water flow rate eventually is turned up to design, the damage will have been done. It is probable that if the design flow rate had been maintained from the onset, the ultimate fouling resistance would have been lower. A similar effect can happen if the cooling water inlet temperature falls below the design figure and the flow rate is again turned down.

Six Types of Fouling

Generally speaking, the types of fouling experienced in most CPI operations can be divided into six fairly distinct categories. First is crystallization — the most common type of fouling which occurs in many process streams, particularly cooling tower water. Frequently superimposed with crystallization is sedimentation which usually is caused by deposits of particulate matter such as clay, sand or rust. From chemical reaction and polymerization often comes a build up of organic products and polymers. The surface temperature and presence of reactants, particularly oxygen can have a very significant effect. Coking occurs on high temperature surfaces and is the result of hydrocarbon deposits. Organic material growth usually is superimposed with crystallization and sedimentation and is common to sea water systems. And corrosion of the heat transfer surface itself produces an added thermal resistance as well as a surface roughness.

In the design of the plate heat exchanger, fouling due to coking is of no significance since the unit cannot be used at such high temperatures. Corrosion also is irrelevant since the metals used in these units are non-corrosive. The other four types of fouling, however, are most important. With certain fluids such as cooling tower water, fouling can result from a combination of crystallization, sedimentation and organic material growth.

A Function of Time

From Figure 9.6, it is apparent that the fouling process is time dependent with zero fouling initially. The fouling then builds up quite rapidly and in most cases, levels off at a certain time to an asymptotic value as represented by curve A. At this point, the rate of deposition is equal to that of removal. Not all fouling levels off, however, and curve B shows that at a certain time the exchanger would have to be taken off line for cleaning. It should be noted that a PHE is a particularly useful exchanger for this type of duty because of the ease of access to the plates and the simplicity of cleaning.

In the case of crystallization and suspended solid fouling, the process usually is of the type A. However, when the fouling is of the crystallization type with a pure compound crystallizing out, the fouling approaches type B and the equipment must be cleaned at frequent intervals. In one particularly severe fouling application, three PHEs are on a 4½ hour cycle and the units are cleaned in place for 1½ hours in each cycle.

Biological growth can present a potentially hazardous fouling since it can provide a more sticky type surface with which to bond other foulants. In many cases, however, treatment of the fluid can reduce the amount of biological growth. The use of germicides or poisons to kill bacteria can help.

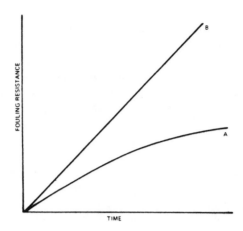

Figure 9.6 Buildup of fouling resistance.

Lower Resistance

It generally is considered that resistance due to fouling is lower with plate heat exchangers than with tubular units. This is the result of five PHE advantages:

1. There is a high degree of turbulence which increases the rate of foulant removal and results in a lower asymptotic value of fouling resistance.
2. The velocity profile across a plate is good. There are no zones of low velocity compared with certain areas on the shell side of tubular exchangers.
3. Corrosion is maintained at an absolute minimum.
4. A smooth heat transfer surface can be obtained. If necessary, the plate surface can be electropolished.
5. In certain cooling duties using water to cool organics, the very high water film coefficient maintains a moderately low metal surface temperature which helps prevent crystallization growth of the inverse solubility compounds.

The most important of these is turbulence. HTRI (Heat Transfer Research Incorporated) has shown that for tubular heat exchangers, fouling is a function of flow velocity and friction factor. Although flow velocities are low with the plate heat exchanger, friction factors are very high and this results in lower fouling resistance.

Marriot of Alfa Laval has produced a table showing values of fouling for a number of plate heat exchanger duties (Table 9.6). These probably represent about one half to one fifth of the figures used for tubulars as quoted in Kern but it must be noted that the Kern figures probably are conservative, even for tubular exchangers.

APV, meanwhile has carried out test work which tends to confirm that fouling varies for different plates with the more turbulent type of plate providing the lower

Table 9.6.

Fluid	Fouling Resistance $(BTU/hr\,ft^2\,{}^\circ F)^{-1}$
Water	
Demineralized or distilled	.00005
Soft	.00010
Hard	.00025
Cooling Tower (treated)	.00020
Sea (coastal) or estuary	.00025
Sea (ocean)	.00015
River, canal	.00025
Engine jacket	.00030
Oils, lubricating	.00010 - .00025
Oils, vegetable	.00010 - .00030
Solvent, organic	.00005 - .00015
Steam	.00005
Process fluid, general	.00005 - .00030

fouling resistances. In testing an R405 heating a multi-component aqueous solution containing inverse solubility salts (i.e. salts whose solubility in water decreases with increasing temperature), it was learned that the rate of fouling in the Paraflow was substantially less than that inside the tubes of a tubular exchanger. The tubular unit had to be cleaned every 3 or 4 days while the Paraflow required cleaning about once a month. Additional proprietary studies have been conducted for APV by HTRI (1) to obtain comparative data on fouling with cooling tower water using an R405 Paraflow and tubular units.

Conclusion

It has been shown that quoting a high fouling resistance can negate a plate heat exchanger design by adding large amounts of surface and thereby overriding the benefits of the high coefficients. Fouling design resistance, therefore, should be chosen with care, keeping in mind that with a PHE it always is possible to add or subtract surface to meet exact fouling conditions.

CORROSION AND HEAT TRANSFER

Design Features

The design principles, methods of construction and the operating characteristics of the plate heat exchanger are fundamentally different from the more traditional forms of heat transfer equipment such as the shell and tube heat exchanger. These have to be taken into account when specifying materials of construction.

The plate heat exchanger consists of a series of gasketed metal plates compressed together within a sturdy frame. The heating/cooling medium and process stream inlets and outlets are at the four corners of the plate and the desired flow pattern through the plate pack is achieved jointly by positioning peripheral elastomeric gaskets to shut off flow paths and by the use of blanking discs to block ports of individual plates.

Plate Pattern

The plates are made from thin gauge material, the thickness of which varies between 0.048 in^2 and 0.024 in^2, formed into a troughed or chevron pattern. The plate pattern has two functions: to induce turbulence in the process stream and thus achieve a high heat transfer coefficient, and to provide reinforcement and plate support points to maintain interplate separation. The large number of plate-to-plate contact points which results when the plates are formed into a pack (especially with the chevron pattern) produces full plate support on a relatively fine matrix. Thus it is possible to achieve high operating pressures (up to 300 psig) with very thin gauge material.

The configuration and geometry of the heat transfer surface is clearly different from that of a shell and tube heat exchanger and frequently, a new materials specification philosophy has to be adopted when considering materials of construction for a machine to operate on a particular process stream.

Combatting Erosion Corrosion

As good quality water supplies become scarcer and more expensive, chemical engineers are looking to alternative sources for cooling purposes. By far the largest reserve available to plants located on the coast or on tidal rivers is sea water. In view of the general corrosivity of such water supplies, however, some of the larger processing complexes are installing sweet water circuits, i.e., a large heat exchanger operating between sea water and a reservoir of good quality water which is circulated through the installation for all cooling duties. The advantage of this system is that it requires only one piece of equipment in corrosion resistant material, the rest of the plant being constructed in carbon steels or alloys most suitable for the process streams. There is no added complication of having to cater for corrosion resistance against the cooling water.

Conventionally, the copper based alloys such as aluminum brass or 90/10 and 70/30 copper nickel alloys have been specified for shell and tube heat exchangers. These materials provide excellent service life. However, successful as these alloys are, their susceptibility to erosion corrosion is well known. The presence of marine growth within the tube of the heat exchanger has disastrous results insofar as it induces turbulence and causes a classical horseshoe shaped erosion pattern to develop. In a relatively short period of time, this results in tube perforation.

Much has been written about critical sea water velocities, and maximum design velocities for the various copper based alloys are well established. In spite of this, erosion corrosion is encountered in a poorly designed shell and tube heat exchanger, especially at the entry to the tubes. Sometimes it is necessary to fit plastic inserts to overcome this problem. These so-called "entry conditions" with their characteristically high level of turbulence result from a change of direction of water flow from the water box into the tube. With the plate heat exchanger, the geometric form of the plate is specifically designed to produce a high level of turbulence and in effect, "entry conditions" prevail over the entire heat transfer surface.

It is quite clear that materials which have been successfully employed for the construction of tubular units to work on sea and brackish water duties need not necessarily give erosion free service in the plate heat exchanger. It is advisable, therefore, when specifying materials for these types of duties to consider materials which will combine excellent corrosion resistance to chloride ions with immunity to erosion corrosion. Materials such as Monel or titanium are the obvious answer although they are more expensive on a pound for pound basis. When the costing is related to a particular thermal duty, however, a plate heat exchanger in these more expensive alloys frequently is less expensive than a shell and tube heat exchanger in the conventional metals.

Corrosive Liquids and Elastomers

One of the obvious differences between a tubular and a plate heat exchanger is the use of plate gaskets in the latter type of unit. Each individual plate in PHE carries a molded gasket positioned in a channel that extends around the corner ports and along the plate edges. This gasket is offered in various materials and is securely bonded in place to effect an overall seal that will prevent the escape of liquids from the heat transfer surface and eliminate any possibility of the intermixing of product and service fluids. Furthermore, it is used to block off the flow path of one of the process streams in order to direct liquids into the desired processing pattern.

The prime requirement for a plate heat exchanger gasket is that it must provide a long lasting seal of high integrity. To achieve this, two properties are of paramount importance — thermal stability and resistance to assuming permanent deformation. The gasket must have good ageing characteristics and low compression set. All the rubber formulations developed for PHE gasketing materials are designed with these as the main criteria.

Gasket Materials

Although non-elastomeric materials such as compressed asbestos fibre have been employed in plate heat exchangers, it is advantageous wherever possible to specify an elastomeric material for gasket use. There are many elastomeric materials available from which gaskets can be molded, each with attributes suitable for specific applications. After extensive work on the optimization of formulation, however, it has been established that the bulk of duties in which a rubber gasket can be employed can be done with five polymer types. These are SBR (styrene butadiene rubber), Medium Nitrile (acrylonitrile butadiene rubber), Butyl (a co-polymer of isobutylene and isoprene), Silicone, or Fluoroelastomer (Viton). The recent introduction of EPDM (ethylene propylene diene methylene) will however supersede the use of butyl while offering somewhat better thermal stability.

Elastomer Characteristics

As might be expected, various elastomeric materials developed for gasket use have fairly well defined advantages and limitations. SBR is suitable for general purpose applications involving aqueous systems at temperatures up to 180°F. Medium nitrile combines the general purpose characteristics of SBR with an excellent fat (and aliphatic hydrocarbons) resistance and an upper working temperature of 285°F. Butyl (or EPDM) exhibits excellent chemical resistance to a wide range of chemical environments such as acids, alkalis, some ketones, and amines although it has poor fat resistance. Silicone is used in a limited number of applications but is the rubber "par excellence" for sodium hypochlorite and general low temperature use. Fluoroelastomers, being expensive, have a somewhat restricted use but have no equal for high temperature (300°F+) applications in oil. Furthermore, by correct formulation, they can be used in 98% sulfuric acid at temperatures up to 212°F. The original disadvantage of this type of elastomer, namely its poor resistance to high temperature steam, has largely been overcome by the introduction of a new range of products.

Parameters of Gasket Selection

Like plate materials, gasket selection is based on a number of inter-related factors, the chemical composition of the process stream and operating temperatures being the most important. In a heating duty, it is necessary to consider the highest temperature prevailing in the machine, i.e., the inlet temperature of the steam or hot water, when considering the gasket material. In some cases where only very high pressure steam in excess of 60 psig or superheated steam is available, it may be necessary to reduce this to a somewhat lower value temperature. Again, this is not a detrimental feature since because of the very high heat transfer coefficients achieved, the need for a high temperature heating medium is obviated. Where this is not a feasible proposition, it may be necessary to specify compressed asbestos fibre gaskets which raise the upper working temperature in the plate heat exchanger to 500°F.

As is so often the case with plate material failures, gasket failures are most frequently attributable to inadequate definition of the process stream. Two clear examples can be cited. The first occurred when butyl rubber was specified for gaskets where the heating medium was steam containing acetic and lactic acid as impurities. After only six months of operation, the gaskets showed signs of deterioration. Upon investigation, it was shown that the steam also contained approximately 50 ppm of high molecular weight fatty acids (stearic/oleic/linoleic) which were condensing out as a solid phase on to the plate and gaskets. Subsequent absorption by the rubber caused softening, loss of strength and general breakup. The problem was resolved by employing a specially formulated medium nitrile rubber which had adequate resistance to both the low and high molecular weight acids.

The other example was where a processor stated that the heating medium was steam at 22 psig (262°F) and gasket failure occurred within six weeks. An investigation confirmed the steam to be at 22 psig but also revealed that it contained over 100°F of superheat. This problem was successfully handled by the installation of a simple desuperheating spray in the steam line.

TYPICAL APPLICATIONS

Closed Circuit Cooling

As the term implies, closed circuit cooling is the utilization of heat transfer equipment to permit the reuse of clean in-plant water. In many cases, this is done by employing a heat exchanger on what might be termed a 'one-on-one' basis, i.e., a single plate heat exchanger is tied in with a single piece of plant operating equipment.

Typically, this arrangement is used in a sulfuric acid plant built as an ancillary facility to a copper smelting operation. With the plant located in the arid hill country of southwestern New Mexico, water conservation was necessary so two Series R145 Paraflows were installed as weak acid coolers. These units are used in conjunction with wet hydraulic scrubbing systems to clean converter off-gas, and in the process, recycle large quantities of scrubber water. The Paraflows handle 1.7 million pounds/hour of a 1% sulfuric acid solution, cooling the liquor from 120°F to 85°F with 72°F cooling water. Heat picked up by the scrubber liquor is removed by passing the stream through the heat exchangers with only a small amount of make-up cooling tower water added to the circuit as the clean cooling water is repeatedly cycled through the scrubbing system.

A simple closed circuit arrangement with a slightly different objective and operation currently is being used by a New England manufacturer of molded wood pulp products. The problem here was to dissipate large amounts of excess heat from a groundwood process stream and thereby prevent discoloration of the pulp. While standard techniques of overflowing sufficient high solids process water to the municipal sewer system reduced temperatures in the white water tank, costly heat was lost in the overflow stream and new problems of thermal and bulk fiber pollution occurred. The solution was to install an APV Series R56 Paraflow in a closed circuit system. This unit handles a heat load of about 6,000,000 BTU/hr while cooling 250 GPM of hot process water to 140°F with 230 GPM of 61°F city water. The cooled white water is discharged to a white water tank, the 140°F process water is recycled and the heated water from the Paraflow is used for other mill heating duties.

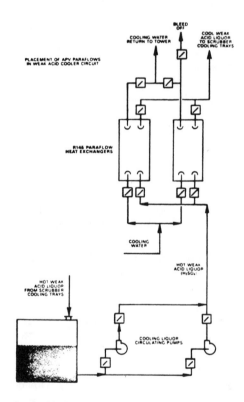

Figure 9.7 Closed circuit cooler in sulfuric plant.

Sweet Water Cooling

Often used in both medium and large capacity plants which draw cooling water from nearby rivers or the sea, sweet water cooling provides the advantage of collecting all risks of fouling and corrosion associated with polluted or saline coolants into one relatively small area. This system utilizes heat transfer equipment to remove heat from clean cooling water so that the clean or "sweet" water can be recirculated as a coolant through various pieces of plant equipment without plugging or corrosion problems. General applications include chemical plants requiring both liquid cooling and condensing duties at various parts of the process, municipal or industrial power generating stations where a large supply of cooling water is needed for auxiliary equipment, offshore oil production platforms where a closed circuit glycol/sea water mixture is used to cool compressed gas and lubricating oil, and iron and steel mills where enormous amounts of heat must be dissipated or redirected from furnace walls, lubricating oils, water jackets and other hot spots.

Fouling Duty

As every plant engineer knows, one of the major aggravations in processing chemicals is the product that literally gums up the works every few weeks.

That is the case in one Glidden Durkee Baltimore plant where the cooling of titanium dioxide slurry causes extreme fouling in conventional shell and tube heat exchangers. Two shell and tube units with 361 tubes apiece have to be shut down on the average of 4–8 times per year each to take care of plugged tubes. Since the 12' long tubes must be degunked individually with high pressure water and cleaning requires approximately eight hours per unit, annual downtime totals more than 200 hours. In addition, erosion and corrosion over a period of time has forced permanent corking of some tubes and a further cut in capacity.

The contrast in the company's newer Baltimore plant is startling. Here two R56 Paraflow plate heat exchangers have been in operation for many months. These APV units are easily meeting requirements of cooling 220,000 lbs/hr of 156°F water to 120°F. No excessive pressure drop due to fouling has been experienced to date. They have proved their ability to handle the TiO_2 slurry without trouble and have never been opened for cleaning. Furthermore, the compact 3'8" × 2'6" Paraflows take only a fraction of the floor space needed for the old shell and tube exchangers and, if and when maintenance is required, permit full access to all plate surfaces for quick cleaning or inspection.

APPLICATIONS ON SODA ASH PROCESSING

Caught between a growing demand for soda ash and an aging production facility in Syracuse, New York, Allied Chemical Corporation decided to spend the money necessary to return this plant to its rated capacity. Approximately 20% was spent in upgrading the heat exchange areas of this operation. The Series R10 APV Paraflow plate heat exchanger was chosen as the optimum replacement for shell and tube units then in use.

Figure 9.8 Two large Series R145 APV Paraflows help conserve water through closed circuit arrangement.

Figure 9.9 R56 Paraflows cool 220,000 lbs/hr of TiO$_2$ slurry from 156–120° F.

Multi-Step Process

Solvay soda ash is a white anhydrous powdered or granular material containing well above 99% sodium carbonate (Na_2CO_3) when shipped to Allied's customers for use in making glass, paper, soap and other chemicals.

Tracing a somewhat complicated process in the simplest of terms, clarified brine first is passed to absorption columns where it absorbs ammonia and carbon dioxide. The brine then is cooled and pumped to a series of carbonating towers to precipitate out $NaHCO_3$ is charged to calciners and dried to Na_2CO_3 soda ash. Ammonia used in the process, meanwhile, is recovered by distillation.

Figure 9.10 Compact R10 Paraflow exchanges 16,000,000 BTUs/hr while two large tubular units await the dismantling crew.

Heat of Reaction

Since every step in the operation produces heat in significant quantities, one of the keys to more efficient production is more efficient cooling of process liquids. Under Allied's original setup, large shell and tube exchangers were used for cooling duties in three locations along the process line.

In the ammonia absorption area, twelve heat exchangers are required to cool circulating liquor from four ammonia absorption columns. Two exchangers handle the initial cooling duty for each column and two additional units are used as aftercoolers before liquor is fed to carbonating columns. The remaining two units are held in reserve for use during cleaning and maintenance downtime periods. The shell and tube units are 6′ in diameter by 18′ long, contain 776 aluminum tubes, provide 3050 sq ft of heat transfer surface and occupy 108 sq ft of floor space.

At the time of this report, five R10 Paraflows were installed in this area. They are equipped with commercially pure titanium plates to resist corrosion, provide 890 sq ft of effective heat transfer surface, and require only about 20 sq ft of floor space. The R10 Paraflows have reduced cleaning time by ½ or 200 man hours per year per unit and have reduced water consumption. When conversion is completed, cooling the ammonia absorption area will not be a production bottleneck.

CBR Coolers

Attention next was focused on replacing four 5′ × 13′ tubular exchangers at the CBR stations because of increasing tube failure and the need for frequent retubing. These units are used to cool the discharge liquor from carbonating towers to the desired feed temperature for the precipitating columns. Again, titanium plated Series R10 Paraflows were selected after it was determined that two plate heat exchangers could handle the duty of the four tubular machines. Each Paraflow contains 33 plates offering a total of 364.8 sq ft of heat transfer surface for cooling 1500 GPM of liquor from 107°F to 100°F. Cooling water consumption is significantly reduced and once more maintenance time is cut by about 50%.

Again, Two for One

Finally, two additional R10 Paraflows were installed to replace four tubular exchangers that had proved to be undersized and inefficient in cooling lubricating water for Nordberg engines. Each R10 cools 750 gal/min of a dilute soda ash solution that is injected into the gas compressor cylinders and has experienced no difficulty in maintaining the desired gas temperature necessary to help meet soda ash production targets.

Substituting the APV plate heat exchangers has greatly reduced compressor gas valve maintenance, increased on-stream time and saved 200 man hours per year of cleaning labor.

NOMENCLATURE

a	constant
A	constant
B	constant
c	constant
Cp	specific heat
d	hydraulic mean diameter ft
h	film heat transfer coefficient BTU/hr ft^2 °F
k	thermal conductivity BTU/hr ft^2 °F
L	length ft

m   constant

n   constant

r   resistance to heat transfer $(BTU/hr\,ft^2\,°F)^{-1}$

s   heat transfer area ft^2

U   overall heat transfer coefficient $BTU/hr\,ft^2\,°F$

V   velocity ft/sec

x   constant

y   constant

Nu   Nusselt number hd/k

Pr   Prandtl number $Cp.\,\mu/k$

Re $Vd\rho/\mu$

ρ density lb/ft^3

μ viscosity lb/ft sec

ΔT temperature difference deg F

REFERENCES

1. Cooper, A., J. W. Suitor and J. D. Usher, Cooling Water Fouling in Plate Heat Exchangers — Sixth International Heat Transfer Conference Toronto (1978).

CHAPTER 10

SPIRAL FLOW HEAT EXCHANGERS

PAUL E. MINTON
Union Carbide Corporation
South Charleston, W. VA

Two types of heat exchangers are constructed such that they present a spiral flow path for one or both fluids being handled. One, the spiral-tube exchanger, is a modification of the shell-and-tube exchanger. The other, the spiral-plate exchanger, is a type of plate exchanger. These two types of heat exchangers therefore present a curved fluid-flow path which offers several advantages over conventional shell-and-tube heat exchangers: compactness; centrifugal forces increase heat transfer; the compact configuration results in a shorter undisturbed flow length; resistance to fouling; relative ease of cleaning; the spiral configuration reduces stresses associated with differential thermal expansion. These curved-flow-path units are particularly useful for handling viscous fluids, slurries, or sludges. This is especially true for the spiral-plate exchanger because of the single flow passage.

Spiral-tube and spiral-plate heat exchangers are generally more expensive than shell-and-tube exchangers having the same heat-transfer surface. Better heat transfer, compactness, and lower maintenance costs often make these spiral exchangers more economical choices than the shell-and-tube exchanger.

The details of the velocity profile in a curved flow path are not completely known. Most investigators agree that there is a secondary flow in curved paths, which is caused by the action of centrifugal force. The extra energy absorbed in this secondary flow, which causes fluid mixing, increases pressure drop and lowers the resistance to heat transfer. Because the nature of secondary flow and the distorted mean-velocity profile are not known, verified theoretical models for pressure drop and heat transfer in curved flow geometries have not been developed. Empirical equations have been developed which predict the results of the majority of past investigators as well as tests on operating spiral heat exchangers.

When these exchangers are arranged to permit both fluids to flow in spiral flow paths, the flow can be essentially countercurrent. The flow is not strictly countercurrent because, throughout the unit, each flow channel is adjoined by an ascending and a descending turn of the other channel. Also, the heat-transfer areas are not equal for each side of the channel because of changing diameters of curvature. A correction factor for the mean temperature difference can be applied; however, it is so small it can generally be ignored.

Although the spiral-plate and spiral-tube exchangers are similar, their methods of fabrication are quite different. The application of each may also be different.

Spiral-plate exchangers are available from a single supplier, as are spiral-tube exchangers. The two types of exchangers are manufactured by two independent companies, however.

SPIRAL-PLATE EXCHANGERS

Fabrication

A spiral-plate exchanger is fabricated from two relatively long strips of plate, which are spaced apart and wound around an open, split center to form a pair of concentric spiral passages. Spacing is maintained uniformly along the length of the spiral by spacer studs welded to the plate.

For most services, both fluid-flow channels are closed by welding alternate channels at both sides of the spiral plate (Figure 10.1). In some applications, one of the channels is left completely open on both ends and the other closed at both sides of the plate (Figure 10.2). These two types of construction prevent the fluids from mixing.

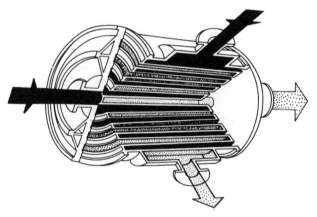

Figure 10.1 Spiral flow in both channels is widely used. Sealing is effected by welding alternate channels on each end of the plate.

Spiral-plate exchangers are fabricated from any material that can be cold worked and welded. Materials commonly used include: carbon steel, stainless steels, nickel and nickel alloys, titanium, Hastelloys, and copper alloys. Baked phenolic-resin coatings are cometimes applied. Electrodes can also be wound into the assembly to anodically protect surfaces against corrosion.

Spiral-plate exchangers are normally designed for the full pressure of each passage. The maximum design pressure is 150 psi because the turns of the spiral are of relatively large diameter, each turn must contain its design pressure, and plate

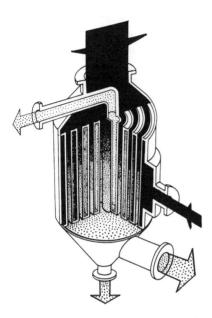

Figure 10.2 Flow is spiral in one channel, axial in the other. Sealing is effected by welding the spiral flow channel on both sides of the plate.

thicknesses are somewhat limited. For smaller diameters, however, the design pressure may sometimes be higher. Limitations of the material of construction govern design temperatures.

The spiral assembly can be fitted with covers to provide three flow patterns:

1. both fluids in spiral flow,
2. one fluid in spiral flow and the other in axial flow across the spiral, and
3. one fluid in spiral flow and the other in a combination of axial and spiral flow.

Flow Arrangements and Applications

For spiral flow in both channels, the spiral assembly includes flat covers at both sides (Figure 10.1). In this arrangement, the fluids usually flow countercurrently, with the cold fluid entering at the periphery and flowing toward the core, and the hot fluid entering at the core and flowing toward the periphery. For this arrangement, the exchanger can be mounted with the axis of the spiral either vertical or horizontal. This arrangement finds wide application in liquid-to-liquid service, and for gases or condensing vapors if the volumes are not too large for the maximum flow area of 72 square inches.

For spiral flow in one channel and axial flow in the other, the spiral assembly includes conical covers, dished heads, or extensions with flat covers (Figure 10.2). In this arrangement, the passage for axial flow is open on both sides and the spiral flow channel is sealed by welding on both sides of the plate. Exchangers with this arrangement are suitable for services in which there is a large difference in the volumes of the two fluids. This includes liquid-liquid service, heating or cooling gases, condensing vapors, or boiling liquids. Fabrication can provide for single pass or multipass on the axial-flow side. This arrangement can be mounted with the axis of the spiral either vertical or horizontal. It is usually vertical for condensing or boiling.

For combination flow, a conical cover distributes the axial fluid to its passage (Figure 10.3). Part of the open spiral is closed at the top, and the entering fluid flows only through the center part of the assembly. A flat cover at the bottom forces the fluid to flow spirally before leaving the exchanger. This type is most often used for condensing vapors and is mounted vertically. Vapors flow first

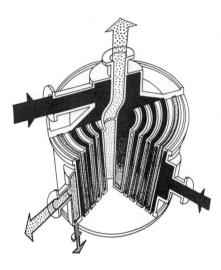

Figure 10.3 Combination flow is used to condense vapors.

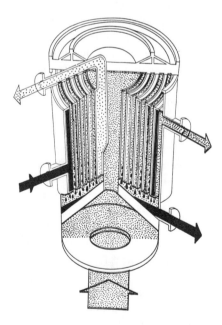

Figure 10.4 Modified combination flow is column mounted.

axially until the flow volume is reduced sufficiently for final condensing and sub-cooling in spiral flow.

A modification of combination flow is the column-mounted condenser (Figure 10.4). A bottom extension is flanged to mate the column flange. Vapor flows upward through a large central tube and then axially across the spiral, where it is condensed. Subcooling may be achieved by falling-film cooling or by controlling a level of condensate in the channel. In the latter case, the vent stream leaves in spiral flow and is further cooled. The column mounted condenser can also be designed for updraft operation and allows condensate to drop into an accumulator with a minimum amount of subcooling.

Advantages

The spiral-plate exchanger offers many advantages over the shell-and-tube exchanger:

1. Single flow passages make it ideal for cooling or heating slurries or sludges. Slurries can be processed in the spiral-plate at velocities as low as 2 feet per second. For some sizes and design pressures, eliminating or minimizing the spacer studs enable this exchanger to handle liquids having a high content of fibrous materials.

2. Fluid distribution is good because of the single flow passage.

3. The spiral-plate exchanger generally fouls at much lower rates than the shell-and-tube exchanger because of the single flow passage and the curved flow path. If fouling does occur, the spiral-plate can be effectively cleaned chemically because of the single flow path. Because the spiral can be fabricated with identical flow passages for the two fluids, it is used for services in which the switching of fluids

allows one fluid to remove the fouling deposited by the other. The maximum plate width of six feet and alignment of spacer studs permit the spiral-plate to be easily cleaned with high-pressure water or steam.

4. The spiral-plate is well suited for heating or cooling viscous fluids because its length to diameter (L/D) ratio is lower than that of straight tubes or channels. Consequently, laminar-flow heat transfer is much higher for spiral plates. When heating or cooling a viscous fluid, the spiral should be oriented with the axis horizontal. With the axis vertical, the viscous fluid stratifies and the heat transfer is reduced as much as 50 percent.

5. With both fluids flowing spirally, countercurrent flow and long passage lengths enable close temperature approaches and precise temperature control. Spiral-plates frequently can achieve heat recovery in a single unit which would require several tubular exchangers in series.

6. Spiral-plates avoid problems associated with differential thermal expansion in non-cyclic services.

7. In axial flow, a large flow area affords a low pressure drop and can be of especial advantage when condensing under vacuum or when used as a thermosiphon reboiler.

8. The spiral-plate exchanger is compact: 2000 square feet of heat transfer surface can be obtained in a unit 58 inches in diameter with a 72 inch wide plate.

Limitations

There are certain disadvantages associated with the spiral-plate heat exchanger:

1. The maximum design pressure is 150 psi, except for some limited sizes.

2. Repair in the field is difficult. A leak cannot be plugged as in a shell-and-tube exchanger. However, the possibility of leakage is much less in the spiral-plate because it is fabricated from plate generally much thicker than tube walls and stresses associated with thermal expansion are virtually eliminated. Should a spiral-plate need repairing, removing the covers exposes most of the welding of the spiral assembly. Repairs on the inner parts of the plates are complicated, however.

3. The spiral-plate exchanger is sometimes precluded from service in which thermal cycling is frequent. When used in cycling services, the mechanical design must be altered to provide for the higher stresses associated with cyclic services. Full-faced gaskets of compressed asbestos are not generally acceptable for cyclic services because the growth of the spiral plates cuts the gasket, which results in excessive fluid by-passing and, in some cases, erosion of the cover. Metal-to-metal seals are generally necessary when frequent thermal cycling is expected.

4. The spiral-plate exchanger usually should not be used when a hard deposit forms during operation, because the spacer studs prevent such deposits from being easily removed by mechanical cleaning. When, as for some pressures, spacer studs can be omitted, this limitation is not present.

5. For spiral-axial flow, the flow paths are not countercurrent and temperature differences must be corrected. The conventional correction for cross-flow applies. Fluids are not mixed; flows are generally single pass. Axial flow may be multipass.

6. Spiral flow arrangements mounted with the axis of the spiral horizontal do not permit fluids to drain under gravity unless drain connections are provided at the low point of each turn of the spiral. Such drains may be of small size, and it may be necessary to remove fluids by purging with air, nitrogen, or some other suitable gas.

Spiral-Plate Design Standards

Table 10.1 presents design standards for spiral-plate heat exchangers. The diameter of the outside spiral can be calculated using the equation below.

$$D_s = [15.36\ L\ (d_c + d_h + 2t) + C^2]^{1/2}$$

D_s = spiral diameter, inches
d_h = channel spacing of hot side, inches
d_c = channel spacing of cold side, inches
t = plate thickness, inches
C = core diameter, inches
L = plate length, feet

Table 10.1. Some Spiral-Plate Exchanger Standards.

Plate Widths, Inches	Outside Diameter Maximum, Inches	Core Diameter, Inches
4	32	8
6	32	8
12	32	8
12	58	12
18	32	8
18	58	12
24	32	8
24	58	12
30	58	12
36	58	12
48	58	12
60	58	12
72	58	12

Channel spacings, inches:
    3/16 (12 inches maximum width), 1/4 (48 inches maximum width), 5/16, 3/8, 1/2, 5/8, 3/4, 1
Plate thicknesses: stainless steel, 14 to 3 U.S. gage:
    Carbon steel, 1/8, 3/16, 1/4, 5/16 inches

Thermal Design Considerations

Empirical equations can be used to predict heat transfer performance and pressure drop of spiral-plate exchangers. There are, however, several factors that must be considered when establishing the thermal design of spiral-plate exchangers. A detailed method for design is presented in Reference 1.

Table 10.2 presents a summary of heat transfer and pressure drop equations applicable to spiral-plate exchangers.

Table 10.2. Empirical Equations for Spiral Plate Heat Exchangers.

Mechanism or Restriction	Empirical Equation for Heat Transfer
Spiral Flow	
no phase change, $N_{Re} > N_{Rec}$	$h/cG = (1 + 3.54D_e/D_H)(0.023)(N_{Re})^{-0.2}(Pr)^{-2/3}$
no phase change, $N_{Re} < N_{Rec}$	$h/cG = 1.86(N_{Re})^{-2/3}(Pr)^{-2/3}(D/L)^{1/3}(\mu_f/\mu_b)^{-0.14}$
Spiral or Axial Flow	
condensing vapor, vertical, $N_{Re} < 2100$	$h = 0.925k(g_c\rho^2/\mu\Gamma)^{1/3}$
Axial Flow	
no phase change, $N_{Re} > 10,000$	$h/cG = 0.023(N_{Re})^{-0.2}(Pr)^{-2/3}$
condensing vapor, horizontal, $N_{Re} < 2100$	$h = 0.76k(g_c\rho^2/\mu\Gamma)^{1/3}$
nucleate boiling, vertical	$h/cG = \phi(N_{Re})^{-0.3}(Pr)^{-0.6}(P^2/\rho_L\sigma)^{0.425}$

$\phi = 0.001$ for steel and copper
$\phi = 0.0006$ for stainless steel
$\phi = 0.0004$ for polished surfaces
$G = W\rho_L/A\rho_v$

Mechanism or Restriction	Empirical Equation for Pressure Drop
Spiral Flow	
no phase change, $N_{Re} > N_{Rec}$	$\Delta P = 6.24 \times 10^{-8}\dfrac{L}{\rho}\left[\dfrac{W}{d_sH}\right]^2\left[\dfrac{1.3z^{1/3}}{(d_s + 0.125)}\left(\dfrac{H}{W}\right)^{1/3} + 1.5 + \dfrac{16}{L}\right]$
no phase change, $100 < N_{Re} < N_{Rec}$	$\Delta P = 6.24 \times 10^{-5}\dfrac{L}{\rho}\left[\dfrac{W}{d_sH}\right]\left[\dfrac{1.035z^{1/2}}{(d_s + 0.125)}\left(\dfrac{z_f}{z_b}\right)^{0.17}\left(\dfrac{H}{W}\right)^{1/2} + 1.5 + 16/L\right]$
no phase change, $N_{Re} < 100$	$\Delta P = 4.73 \times 10^{-9}\left[\dfrac{L\rho z}{(d_s)^{2.75}}\right]\left[\dfrac{z_f}{z_b}\right]^{0.17}\left[\dfrac{W}{H}\right]$
Axial Flow	
no phase change, $N_{Re} > 10,000$	$\Delta P = 9.94 \times 10^{-9}\left[\dfrac{W}{L}\right]^{1.8}\left[0.0115z^{0.2}\dfrac{H}{d_s} + 1 + 0.03H\right]$

219

Critical Reynolds Number — For curved flow, the Reynolds number above which turbulent flow is achieved is a function of the degree of curvature. The equation below establishes the critical Reynolds number.

$$N_{Rec} = 20,000 \, (D_e/D_H)^{0.32}$$

N_{Rec} = critical Reynolds number
D_e = equivalent diameter of flow channel
D_H = diameter of a spiral

For a spiral-plate, the value of D_H is not constant for any exchanger.

Length to Diameter Ratio (L/D) — For curved flow paths, the value of L/D is much less than for straight flow paths of the same length. For laminar flow, the value of $(D/L)^{1/3}$ is of importance and can be calculated using the equation below:

$$(D/L)^{1/3} = (d_s/d_H)^{1/6}$$

D   = diameter
L   = length
d_s = channel spacing
d_H = diameter of spiral

Because the spiral diameter and the channel spacing are not constant, the value of $(d_s/d_H)^{1/6}$ varies: the ratio of the high value to the low value of this term is 1.5.

Turbulent Regime Sensible Heat Transfer — For curved flow paths, sensible heat transfer in the turbulent flow regime is greater than that for straight flow paths. For the spiral-plate, sensible heat transfer is increased by the factor $(1 + 3.54 \, D_e/D_H)$ where D_e is the equivalent diameter of the flow path and D_H is the diameter of the spiral. The value of the term $(1 + 3.54 \, D_e/D_H)$ is not constant for any given heat exchanger. A weighted average of 1.1 is frequently used for designing spiral plates.

Condensate Loading for Horizontal Condensation — For a spiral plate, condensate loading depends upon the length of the plate and the spacing between adjacent plates. For any given plate length and channel spacing, the heat-transfer area for each 360-degree winding of the spiral increases with the diameter of the spiral. The number of spiral revolutions affects the condensate loading in two ways: (1) the heat-transfer area changes resulting in more condensate being formed in the outer spirals; and (2) the effective length over which the condensate is formed is determined by the number of revolutions and the plate width. The effective number of spirals can usually be taken as L/7 where L is the plate length in feet. At high condensate loadings, the liquid condensate on the bottom of the spiral channels may blanket part of the effective heat-transfer surface.

Pressure Drops for Spiral Flow Paths — For Reynolds numbers greater than 100, a term (equal to 1.5 in the equations presented) must be applied to account for the pressure drop caused by the spacer studs. The value of 1.5 assumes 18 studs per square foot and a stud diameter of 5/16 inches. For Reynolds numbers less than 100, the spacer studs have little effect on pressure drop and any such effect is included in the equation presented.

Pressure Drop for Axial Flow — The equation for axial flow pressure drop includes the pressure drop resulting from the spacer studs and accounts for pressure losses in inlet and outlet nozzles, provided the nozzles have been properly sized.

SPIRAL-TUBE EXCHANGERS

Spiral-tube exchangers consist of one or more concentric, spirally wound coils clamped between a cover plate and a casing. Both ends of each coil are attached to a manifold fabricated from pipe or bar stock (Figures 10.5 and 10.6). The coils, which are stacked on top of each other, are held together by the cover plate and casing. Spacing is uniformly maintained between each turn of the coil to create a uniform, spiral-flow path for the shellside (casing) fluid.

Figure 10.5 Casing is removed without dis-connecting piping.

Coils can be formed from almost any metallic material of construction; some of the more common materials are carbon steel, copper and copper alloys, stainless

Figure 10.6 Spiral-tube design makes shellside cleaning easy.

steels, and nickel and nickel alloys. Tubes may be finned to provide extended surfaces for heat transfer. Casings are made of cast iron, cast bronzes, carbon steel, and stainless steel.

Tubes may be attached to the manifolds by soldering, brazing, welding, or, in some cases, rolling. Draining or venting can be facilitated by various manifold arrangements and casing connections. Flow through either or both the coil and casing may be single- or multipass; multipass arrangements are obtained by proper baffling.

Standard tubeside pressures are available up to 600 psi, although higher pressures are available. Casing pressures are generally available up to 300 psi; again, higher pressures are available, but sometimes can only be achieved by providing a non-removeable casing. Design temperatures are limited by the materials of constructions and the method used to attach the tubes to the manifolds.

Flow Arrangements and Applications

The tube side flow arrangement must be spiral flow, either single or multipass. The shellside flow arrangement can be spiral flow or axial flow. Shellside spiral flow can be either single or multipass. Axial flow is normally restricted to single pass.

Axial shellside flow is used when condensing or boiling or for high volumetric flow rates. Most applications involve spiral flow both in the coils and in the casing.

Advantages

The spiral-tube exchanger offers several advantages over shell-and-tube heat exchangers:

1. The spiral-tube exchanger is especially suited for low flow rates or small heat loads.
2. The spiral-tube is particularly effective for heating or cooling viscous fluids. It is not as effective as the spiral-plate because of possible mal-distribution as a result of several flow passages.
3. The two fluids flow countercurrently in spiral flow permitting precise temperature control and effective heat recovery. As with the spiral-plate, heat recovery can frequently be achieved in a single unit which would require several shell-and-tube exchangers in series.

4. Spiral-tube units eliminate problems associated with differential thermal expansion.
5. The casing can be easily removed for cleaning or maintenance. Finned tubes can be provided for ease of cleaning.
6. The spiral-tube is compact and easily installed: 325 square feet of heat-transfer surface can be provided in a unit with a diameter of 37 inches and a height of 31 inches. Heat transfer can be increased by fabricating with finned tubes.

Limitations

There are several disadvantages associated with the spiral-tube heat exchanger:

1. The manifolds are relatively small; consequently, repair of leaks at tube-to-manifold joints is difficult. Leaks, however, do not occur frequently, and proper selection of the method of tube attachment can improve reliability.
2. Services are limited to those that do not require mechanical cleaning of the insides of the tubes. The spiral-tube unit can be mechanically cleaned on the shellside, and both sides can be cleaned chemically. Finned tubes can also be used to facilitate shellside cleaning.
3. For some sizes, stainless steel coils must be provided with spacers to maintain uniform shellside flow area. These spacers increase pressure drop and may interfere with mechanical cleaning of the shellside.
4. The maximum flow areas available are restricted.
5. The maximum available heat-transfer surface is restricted in standard units.
6. When the spiral-tube is mounted with the planes of the spiral tubes in a vertical position, the tubeside does not drain under gravity. Drainage can only be accomplished by purging with air, nitrogen, or some other suitable gas.

Spiral-Tube Design Standards

Table 10.3 presents design standards for spiral-tube heat exchangers.

Thermal Design Considerations

As with spiral-plate exchangers, empirical equations can be used to predict heat transfer performance and pressure drop of spiral-tube exchangers. Table 10.4 presents a summary of heat transfer and pressure drop equations applicable to spiral-tube exchangers. There are several factors that must be considered when establishing the thermal design of spiral-tube exchangers. A detailed design method is presented in Reference 2.

Critical Reynolds Number — The criterion for critical Reynolds number discussed for spiral-plate exchangers also applies to spiral-tube exchangers. Again, for spiral-tubes, the value of D_H is not constant for any given unit.

Length to Diameter Ratio (L/D) — The value of $(D/L)^{1/3}$ for the tubeside can be determined from the relation below:

$$(D/L)^{1/3} = (D_i/D_H)^{1/6}$$

D = diameter
L = length
D_i = inside diameter
D_H = spiral diameter

223

Table 10.3. Spiral-Tube Exchanger Design Standards.

No. Tubes	Tube Spacing, In	Shellside Flow Area, Sq In	Standard Lengths, Ft — Tube-side	Standard Lengths, Ft — Shell-side	Heat-Transfer Area, Sq Ft
Tube O.D., 1/4 in					
8	1/8	0.358	4.92	5.7	2.56
12	1/8	0.537	8.11	9.4	6.31
18	3/16	1.08	9.77	11.5	11.5
18	3/16	1.08	15.06	17.5	17.66
30	3/16	1.80	9.77	11.5	19.02
30	3/16	1.80	15.06	17.3	29.5
Tube O.D., 3/8 in					
8	1/8	0.618	5.51	7.9	4.4
12	1/8	0.93	9.9	11.5	11.6
12	1/8	0.93	14.79	15.5	17.4
20	1/8	1.55	9.9	11.5	19.4
20	1/4	2.48	10.98	13.1	21.5
20	5/16	2.95	12.87	15.5	25.2
Tube O.D., 1/2 in					
4	1/8	0.466	5.25	6.5	2.75
6	1/8	0.699	5.25	6.5	4.13
9	1/8	1.04	8.16	9.75	9.63
9	1/8	1.04	10.88	13.0	12.7
15	1/8	1.75	8.16	9.75	16.0
15	3/16	2.23	10.62	12.75	20.9
15	1/4	2.68	12.4	14.9	24.5
15	5/16	3.16	19.25	22.2	37.5
15	5/16	3.16	27.5	30.8	54.0
15	5/16	3.16	33.41	37.2	66.07
15	5/16	3.16	43.2	47.9	84.87
30	1/4	5.37	12.38	14.9	48.9
30	5/16	6.30	19.14	22.2	75.0
30	5/16	6.30	27.5	30.8	108.0
30	5/16	6.30	33.41	37.2	132.15
30	5/16	6.30	43.2	47.9	169.14
Tube O.D., 5/8 in					
12	1/8	1.94	6.6	8.25	13.0
12	3/16	2.42	8.46	11.6	16.6
12	1/4	2.88	12.2	14.5	24.0
12	5/16	3.35	18.13	21.2	35.5
12	5/16	3.35	23.87	27.25	46.8
12	5/16	3.35	29.36	33.4	57.66
12	5/16	3.35	38.05	42.1	74.95
24	1/4	5.76	12.2	14.5	48.0
24	5/16	6.68	18.13	21.2	70.3
24	5/16	6.68	23.87	27.25	93.6
24	5/16	6.68	29.36	33.4	115.32
24	5/16	6.68	38.05	42.1	149.9
Tube O.D., 3/4 in					
10	3/16	2.62	8.12	9.7	15.9
10	1/4	3.09	9.92	12.3	19.4
10	5/16	3.59	15.2	18.25	29.8
10	5/16	3.59	20.61	24.0	40.3
10	5/16	3.59	25.81	29.5	50.76
10	5/16	3.59	33.75	38.0	66.29
10	5/16	3.59	40.3	45.3	79.0
10	5/16	3.59	47.72	53.4	94.0
10	5/16	3.59	55.58	62.2	109.0
20	1/4	6.16	9.87	12.3	39.2
20	5/16	7.10	15.2	18.25	59.6
20	5/16	7.10	20.61	24.0	80.6
20	5/16	7.10	25.81	29.5	101.52
20	5/16	7.10	33.75	38.0	132.58
20	5/16	7.10	40.3	45.3	158.0
20	5/16	7.10	47.72	53.4	188.0
20	5/16	7.10	55.58	62.2	218.0
30	5/16	10.62	33.75	38.0	198.9
30	5/16	10.62	40.3	45.3	237.0
30	5/16	10.62	47.72	53.4	282.0
30	5/16	10.62	55.58	62.2	327.0

Table 10.4. Empirical Equations for Spiral Tube Heat Exchangers.

Mechanism or Restriction	Empirical Equation for Heat Transfer
Tube Side	
no phase change, $N_{Re} > N_{Rec}$	$h/cG = (1 + 3.54 D_e/D_H)(0.023)(N_{Re})^{-0.2}(Pr)^{-2/3}$
no phase change, $N_{Re} < N_{Rec}$	$h/cG = 1.86(N_{Re})^{-2/3}(Pr)^{-2/3}(D/L)^{1/3}(\mu_f/\mu_b)^{-0.14}$
condensing vapor, horizontal, $N_{Re} < 2100$	$h = 0.76k(g_c \rho^2/\mu\Gamma)^{1/3}$
Shell Side	
no phase change, spiral flow, $N_{Re} > N_{Rec}$	$h/cG = (1 + 3.54 D_e/D_H)(0.023)(N_{Re})^{-0.2}(Pr)^{-2/3}$
no phase change, spiral flow, $N_{Re} < N_{Rec}$	$h/cG = 1.86(N_{Re})^{-2/3}(Pr)^{-2/3}(D/L)^{1/3}(\mu_f/\mu_b)^{-0.14}$
condensing vapor, horizontal, axial flow, $N_{Re} < 2100$	$h = 0.76k(g_c \rho^2/\mu\Gamma)^{1/3}$
nucleate boiling, horizontal, axial flow	$h/cG = \phi \, (N_{Re})^{-0.3}(Pr)^{-0.6}(P^2/\rho_L \sigma)^{0.425}$

$$\phi = 0.001 \text{ for steel and copper}$$
$$\phi = 0.0006 \text{ for stainless steel}$$
$$\phi = 0.0004 \text{ for polished surfaces}$$
$$G = W\rho_L/A\rho_v$$

Mechanism or Restriction	Empirical Equation for Pressure Drop
Tube Side	
no phase change, $N_{Re} > N_{Rec}$, single pass	$\Delta P = 4.71 \times 10^{-7} \left[\dfrac{z^{0.15}}{\rho}\right]\left[\dfrac{(L/d_i) + 16}{d_i^{3.75}}\right]\left[\dfrac{W}{n}\right]^{1.85}$
no phase change, $100 < N_{Re} < N_{Rec}$, single pass	$\Delta P = 1.22 \times 10^{-5} \left[\dfrac{W}{n}\right]^{4/3}\left[\dfrac{z^{2/3}}{\rho}\right]\left[\dfrac{z_f}{z_b}\right]^{0.14}\left[\dfrac{L}{d_i^{13/3}}\right]$
no phase change, $N_{Re} < 100$, single pass	$\Delta P = 3.39 \times 10^{-5} \left[\dfrac{W}{n}\right]\left[\dfrac{z}{\rho}\right]\left[\dfrac{L}{d_i^4}\right]$
Shell Side	
no phase change, $N_{Re} > N_{Rec}$, single pass	$\Delta P = 2.67 \times 10^{-7} \left[\dfrac{z^{0.15}}{\rho}\right]\left[\dfrac{W^{1.85}}{a^3}\right] L(nd_o)^{1.15}$
no phase change, $100 < N_{Re} < N_{Rec}$, single pass	$\Delta P = 6.86 \times 10^{-6} \left[\dfrac{z^{2/3}}{\rho}\right]\left[\dfrac{W^{4/3}}{a^3}\right]\left[\dfrac{z_f}{z_b}\right]^{0.14} L(nd_o)^{5/3}$
no phase change, $N_{Re} < 100$, single pass	$\Delta P = 1.92 \times 10^{-5} \left[\dfrac{z}{\rho}\right]\left[\dfrac{W}{a^3}\right] L(nd_o)^2$
Axial flow, $N_{Re} > 1000$, single pass	$\Delta P = 4.18 \times 10^{-10} \left[\dfrac{z^{0.15}}{\rho}\right]\left[\dfrac{W}{L}\right]^{1.85} \dfrac{n\, d_o^{1.41}}{d_s^{3.41}}$

For spiral-tube exchangers, the value of $(D_i/D_H)^{1/6}$ varies: the ratio of the high value to the low value of this term is 1.7.

The value of $(D/L)^{1/3}$ for the shellside can be determined in the same manner as for the spiral plate:

$$(D/L)^{1/3} = (d_s/d_H)^{1/6}$$

The value of $(d_s/d_H)^{1/6}$ varies; the ratio of the high value to the low value of this term is 1.2.

Turbulent Regime Sensible Heat Transfer — As with the spiral plate, heat transfer is increased in the spiral-tube exchanger because of the curved flow path. Weighted averages of the value of the factor $(1 + 3.54 \, D_e/D_H)$ frequently used for designing spiral-tube exchangers are 1.15 for the tubeside and 1.1 for the shellside.

Pressure Drop — No effects of any required spacers have been included in the equations for shellside pressure drop.

NOMENCLATURE, HEAT TRANSFER

A	= heat transfer surface
c	= specific heat
D	= diameter
D_e	= equivalent diameter
D_H	= spiral diameter
g_c	= gravitational constant
G	= mass velocity
h	= film coefficient of heat transfer
k	= thermal conductivity
L	= length
N_{Re}	= Reynolds number
N_{Rec}	= critical Reynolds number
P	= pressure
Pr	= Prandtl number
W	= vapor flow rate
ϕ	= numerical constant
μ	= viscosity
μ_b	= bulk viscosity
μ_f	= film viscosity
Γ	= condensate loading
ρ	= density
ρ_L	= liquid density
ρ_v	= vapor density
σ	= surface tension

NOMENCLATURE, PRESSURE DROP

a	=	net free-flow area, square inches
d_i	=	inside diameter, inches
d_o	=	outside diameter, inches
d_s	=	channel spacing or tube spacing, inches
H	=	plate width, inches
L	=	plate or tube length, feet
n	=	number of tubes
N_{Re}	=	Reynolds number
N_{Rec}	=	critical Reynolds number
ΔP	=	pressure drop, psi
W	=	flow rate, pounds per hour
z	=	viscosity, centipoises
z_b	=	bulk viscosity, centipoises
z_f	=	film viscosity, centipoises
ρ	=	density, pounds per cubic foot

REFERENCES

1. Minton, P.E., "Designing Spiral-Plate Heat Exchangers," *Chemical Engineering*, May 4, 1970, p. 103.
2. Minton, P.E., "Designing Spiral-Tube Heat Exchangers," *Chemical Engineering*, May 18, 1970, p. 145.
3. Lord, R. C., P. E. Minton and R. P. Slusser, "Design of Heat Exchangers," *Chemical Engineering*, January 26, 1970, p. 96.
4. Lord, R. C., P. E. Minton and R. P. Slusser, "Design Parameters for Condensers and Reboilers," *Chemical Engineering*, March 23, 1970, p. 127.
5. Lord, R. C., P. E. Minton and R. P. Slusser, "Guide to Trouble-Free Heat Exchangers," *Chemical Engineering*, June 1, 1970, p. 153.

PROCESS EQUIPMENT SERIES

Volume 2   Index

Rotary screw compressor systems 91

S

T

U

V

W